PRAISE FOR *A BRIEF HISTORY OF THE FEMALE BODY*

"*A Brief History of the Female Body* is a wonderful book giving insights into the biological struggle between mother and child, why you have breasts, orgasms, periods, and menopause. This is a must-read for every woman and those who love her."

—**Louann Brizendine MD**, *New York Times* bestselling author of *The Female Brain* and *The Upgrade*

"In this fascinating and hugely enjoyable book, Dr. Emera explores the deep past to shed light on the present and future of women's health."

—**Lisa Mosconi, PhD**, *New York Times* bestselling author of *The XX Brain*

"In this excellent and engaging book, Emera probes the mysteries and mythology of the female body both through the eyes of an evolutionary biologist and a mother. Her delightful prose and keen insights make for a uniquely enjoyable gynecologic journey of discovery."

—**Jonathan Reisman, MD**, author of *The Unseen Body*

"Understanding ourselves begins with understanding our biology. In this important book, Deena Emera guides us on a journey through the past, present, and future of the female body. It's a story with profound implications not just for women, but for all of us."

—**Jamie Metzl**, author of *Hacking Darwin*

"At last! Now that women like Deena Emera are studying evolutionary biology and medicine, we get to read books like *A Brief History of the Female Body*. Deena provides timely and well-informed summaries and discussion of topics women care about within the spheres that actually matter to us."

—**Sarah Blaffer Hrdy**, author of *Mother Nature*

"Deena Emera generously shares her personal perspectives as a scientist and a woman, telling her story and that of those living in female bodies in a conversational style peppered with humor, making a significant contribution to our understanding of the complexity of the female human body and its evolutionary origins. An educational and enjoyable read!"

—**Nancy Fugate Woods, BSN, PhD, FAAN,** Professor Emerita, University of Washington School of Nursing

"Evolutionary biologist Deena Emera uses both her academic acumen and personal experience to reflect on significant basics of female experience—periods, breasts, orgasms, pregnancy, and menopause. Not only are her reflections valuable to anyone interested in women's health as an important subset of women's studies, but they can literally assist the reader in putting her personal experience into perspective."

—**Angela Barron McBride, PhD, RN**, author and Distinguished Professor Emerita, Indiana University

"In *A Brief History of the Female Body*, Dr. Emera addresses many questions that women raise to better understand their bodies and their health, making this engaging book a must-read for women of all ages, including professionals who care about providing quality health care and understanding of bodily transitions for women."

—**Vivian W. Pinn, MD**, women's health research advocate

A BRIEF HISTORY OF THE

FEMALE BODY

AN EVOLUTIONARY LOOK
at How and Why the Female
Form Came to Be

DEENA EMERA, PhD

Published by Sourcebooks
P.O. Box 4410, Naperville, Illinois 60567–4410
(630) 961-3900
sourcebooks.com

The Library of Congress has cataloged the hardcover edition as follows:
Names: Emera, Deena, author.
Title: A brief history of the female body : an evolutionary look at how
and why the female form came to be / Deena Emera.
Description: Naperville, Illinois : Sourcebooks, [2023] | Includes bibliographical references. |
Summary: "Knowledge is the most powerful weapon. As the female body is constantly being
politicized and policed, it is now more than ever that people must understand the inner
workings of women's body. Written by an evolutionary geneticist, Deena Emera, Ph.D., in an
accessible, nonjudgmental tone, A Brief History of the Female Body unravels misconceptions
women have about their own bodies and supplies evolutionary-backed scientific analysis that
provides a more complete understanding of women's bodies. Covering topics once considered
taboo-from periods, to pregnancy, to the female orgasm-A Brief History of the Female
Body illuminates how the female form has transformed over millions of years to become the
beautiful, unique bodies women see in the mirror each day"-- Provided by publisher.
Identifiers: LCCN 2022055161
Subjects: LCSH: Females--Physiology--Popular works. | Females--Evolution-
-Popular works. | Women--Physiology--Popular works. | Women--
Evolution--Popular works. | Sex differences--Popular works.
Classification: LCC QP81.5 .E455 2023 | DDC 612.6--dc23/eng/20221215

LC record available at https://lccn.loc.gov/2022055161

Printed and bound in the United States of America.
VP 10 9 8 7 6 5 4 3 2 1

For Samir, Alexander, Nile, and Lena

Contents

Author's Note

The idea for this book was conceived during a unique time in my life. While studying the evolution of pregnancy as a graduate student at Yale, I was going through the nine-month journey of my first pregnancy. As I experienced morning sickness and worried about conditions like gestational diabetes and preterm labor, I—like anyone pregnant with their first child—had a thousand questions about the physical and emotional roller coaster I was on. But I was also becoming the biologist who could answer those questions. At many points during my research, I thought to myself, *Others would want to know this. I should write a book.*

Since those glimmers of an idea over ten years ago, I completed my PhD in evolutionary biology, did a postdoc in genetics at the Yale School of Medicine, and am now a scientist in the Center for Reproductive Longevity and Equality at the Buck Institute. (I have also been through three more pregnancies!) I am fascinated with how and why animal bodies evolve, and I have been particularly interested in female biology, investigating the evolution of pregnancy, menstruation, and menopause. These topics excite me both because they showcase the forces and mechanisms that have driven the evolution of animal bodies and because they deepen my knowledge of how and why my own body works the way it does.

This book is for those who also want a deeper understanding of the female body. Rather than writing an encyclopedia or textbook, I have chosen a handful of topics with which I am familiar and fascinated, including pregnancy, menstruation, the mammary gland, orgasms, and menopause. We will discuss how these traits develop and function in women today, but we will mostly focus on *why* they exist in the first place. To understand why, we must go back in time. Most human traits were inherited by ancestors, and depending on the trait, those ancestors might have lived 15,000 years, 3 million years, 100 million years, or even longer ago. We must look to these ancestors to understand female biology today. For traits that may have puzzled or frustrated you, like menstruation and menopause, I will give you the evolutionary context to understand why we live with them. For traits that may seem more obvious, like pregnancy and the mammary gland, I will reveal many twists and surprises that make for great stories.

I hope to present these topics in the most accessible and engaging way that I can. Early in my career, I had a brief stint as a high school biology teacher and became sensitive to the

hurdle that biological jargon imposed on my students' understanding of basic concepts. Even as a scientist familiar with
biological terminology, reading an article outside my immediate field can be excruciating. I do my best in this book to avoid
esoteric language when I can and to explain the jargon when
I can't avoid it. In the introduction, I present the language,
concepts, and theoretical work that provide the foundation
for understanding the rest of the book. The evolutionary logic
might be challenging to grasp at first, but it is also fascinating,
intuitive, and worth the read.

This book focuses on females, which in biology refers to the
sex that produces eggs (as opposed to sperm). Even if we choose
never to use those eggs (or sperm)—a choice only humans have
because of cognitive and technological advances in our species—
our bodies are built for reproduction. Therefore, to understand
human biology, we must talk about sexual reproduction—the
mixing of DNA from two parents to produce children—and
biological sex, as mentioned above. Throughout this book, I
use the terms *female* and *woman* to refer to an individual who
produces eggs, but I recognize that some individuals who are
equipped to make eggs do not identify as a woman or female
(and some equipped to make sperm do not identify as a man
or male). For these people, their biological sex is not the same
as their gender, a social identity that is affected by cultural and
biological factors. We are only beginning to understand how
culture, experience, hormones, and genetics influence gender, a
topic I touch on in chapter I. As I chose the words for this book,

my purpose was not to gloss over or ignore gender but only to streamline the discussion, much of which pertains to the organs and tissues used by female mammals for reproduction.

Lastly, while the idea for this book was conceived during the unusual situation I mentioned above, it was carried to term many years later during another unusual situation—the COVID-19 pandemic. The three years I have spent developing, writing, and editing this book have been difficult ones for most of us in the world. The pandemic has brought illness and death, not to mention stress, disruptions to work and family life, and isolation from others. Writing this book has been a welcome escape in an isolating time, for it has reminded me of the connections we share with all life on earth, and I am grateful for the chance to connect with you through my writing.

INTRODUCTION

The Shaping of Female Biology

I have often puzzled over the idiosyncrasies of female biology. As a teenager, I agonized about my periods, wondering why I had to endure the monthly mood swings, pain, and inconvenience. During my pregnancies, I was baffled by the many possible complications, including miscarriage, high blood pressure, diabetes, premature birth, and postpartum depression. Now in my forties and approaching the end of my reproductive years, I am both mystified and miffed by menopause. Why must I lose my fertility and experience the notorious side effects of menopause while my husband gracefully enters the second half of his life with much less fanfare?

As an evolutionary biologist who has spent years studying the evolution of female reproductive biology, I have searched for the answers to these questions that women everywhere ask about their bodies. My mission in this book is to share with you what I know, hoping to answer your questions too. The aim of this introduction is to give you the tool kit you'll need to understand my explanations in subsequent chapters.

In discussing the evolution of female biology in this book, I will cover the evolutionary process that you've certainly heard about: natural selection. But to better understand the enigmas of the female body, we will explore less familiar evolutionary processes. In the version of evolution often taught in high school and portrayed in the media, we imagine a tooth-and-claw struggle in the jungle between predator and prey, picturing evolution as a result of a battle between animal and nature. But the truth is many of the conflicts our ancestors faced were with *each other*. These evolutionary conflicts of interest—which can exist between male and female over reproduction, between siblings, and even between mother and child—have been an important driving force in the evolution of female biology.

Many of us idealize the mother-child relationship, assuming it is molded by harmonious, shared interests. We are expected to give anything and everything to our children, since what is best for our children should be what is best for us. When mothers prioritize our own needs, we feel guilty and selfish. However, while the relationship between mother and child *is* based on evolved cooperation, perplexingly, it has also been shaped by conflicts of interest that play out in the womb, continue after birth, and even extend into adulthood. A child's evolutionary motivation is to take more from its mother than what is best for her to give: The child tries to get more, and the mother tries to hold her ground. This may sound like a scene from a modern-day living room or kitchen, but in fact it has been going on and on, back and forth, for millions of years. These epic evolutionary conflicts have shaped our monthly periods, our difficult pregnancies, our behaviors with our children, and perhaps even our experience of menopause.

Our ancestors also experienced sexual conflict over whether to have sex and how frequently to do it. Many of our female primate relatives regularly endure rape and sexual intimidation, and in some species, this aggressive male behavior extends to the killing of a female's babies fathered by other males in order to gain sexual access to her. In humans, though, there was a détente between males and females, and coercive male behavior is more the exception than the rule.

So the female story isn't only about conflict. Beyond just an easing of tensions between female and male, the evolution of greater social cooperation in the human lineage has left lasting marks on our anatomy, physiology, and behavior. Early humans evolved a more cooperative family life, with fathers and extended family pitching in to feed and raise children. Not only does this cooperative behavior remain with us today, but it may have changed other aspects of our biology—like the way we age—as you'll learn when we discuss longevity and menopause.

One of the key messages I hope you take from this book is that many features of our biology today can be explained by the evolution of family relationships, with all their cooperation and conflict.

The high school version of evolution

Before diving into female biology, we need to get a handle on some basics. Evolutionary conflict and cooperation both involve Darwin's idea of natural selection, so natural selection is the place we need to start. For a few years after college, I was a tenth grade biology teacher and tried to use the most intuitive and compelling examples to illustrate how adaptation by natural selection works. Some of my go-tos were the different beak shapes of Galápagos finches, color changes in peppered moths during the Industrial Revolution, and the increase in lactose tolerance in humans after the domestication of dairy animals.

Let's use the classic example of peppered moths as a refresher. Moths in Manchester and other parts of England in the early 1800s had light wings with small dark splotches. This color combination is good camouflage against light-colored tree trunks, making it difficult for birds to prey on moths that rest on these trees during the day. But as the Industrial Revolution progressed in the nineteenth century, so much pollution was released from factories that trees became blackened with soot. Importantly, there was some variation in moth wing color as trees started to darken—most moths were light, but a few were dark, and offspring resembled the color of their parents. This kind of inherited variation is necessary for adaptation by natural selection, and now we know it exists in part because mutations are always popping up in our DNA. I can hear my fresh twenty-one-year-old self now, my zealous teacher voice emphasizing the randomness of these mutations. Evolution is not intentional, I would tell my students. Mutations cannot anticipate what will work, but the ones that happen to improve reproductive success increase in frequency. In this case, the camouflaged darker moths fared better than the light ones in the new polluted environment. They weren't eaten by birds and therefore had more babies, passing their dark version of the wing color gene to the next generation. Between about 1850 and 1900, dark moths almost completely replaced light ones.

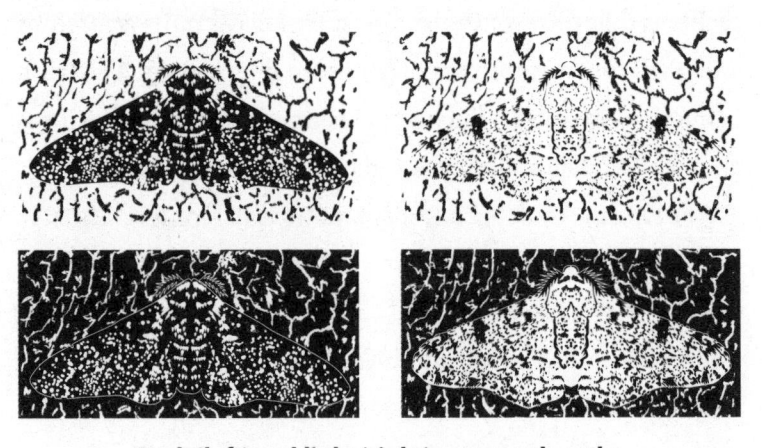

Dark (left) and light (right) peppered moths

As I taught my students about peppered moths, I stressed the importance of environmental challenges in the process of natural selection. Moths faced an obvious obstacle in their environment—their bird predators. While this case focuses on predation as the selective environmental factor, there are plenty of others, including resource availability. Darwin's well-known finches ate different foods on the various Galápagos islands. In a well-studied case from the 1970s, the type of food changed on one of the islands because of a drought. The finches with larger beaks were better able to handle the large, hard seeds, and beak size increased on the island over time.

A more comprehensive view of evolution

While this stripped-down version of evolution was what I could cover in a two-week high school unit on the subject, the reality is that the evolution of most biological traits is rarely as straightforward as the case of peppered moth wings. The truth is much more interesting. Not every body part or behavior is an obvious adaptation produced by natural selection.

First, some traits are not adaptations at all. Traits can emerge simply as by-products of others, the most obvious example of which is the belly button. The belly button has no function, and it didn't evolve as an adaptation for any specific function; it is simply a by-product of the placenta and umbilical cord evolving in placental mammals as an alternative way to nourish developing babies. Another example is nipples in males. In a controversial and influential paper in 1979, Stephen Jay Gould and Richard Lewontin called these biological by-products "spandrels" after the architectural term for the leftover triangular spaces in structures built with arches. Architectural spandrels weren't designed to fulfill a specific function; they are simply features that arise because of the way structures with arches are built. As Gould and Lewontin asserted, many biological traits are like spandrels, arising as nonadaptive by-products of the evolution of other traits that are adaptive. In a later chapter, I will argue that menstruation in our species is a by-product trait (albeit with a fascinating backstory).

Gould and his colleagues stressed that originating as a spandrel does not prevent a trait from becoming useful later. Like architectural spandrels, which are often beautifully decorated to enhance the aesthetics of structures, biological spandrels can become useful and contribute to an organism's success, even if it didn't emerge for that (or any) functional purpose. Gould used the example of the human brain, arguing that when the large, complex brain evolved for its main function (perhaps consciousness, although he stressed that we do not really know), thousands of spandrels emerged, such as the capacities for language, reading, writing, art, commerce, and religion. Gould saw many aspects of human mental uniqueness not as adaptations but simply as by-products of a complex brain, and he used this example to support his more general (and still contentious) claim that spandrels are a frequent and important source of evolutionary innovation. In chapter 9, I'll raise the possibility that human menopause emerged as a by-product of longer life spans in humans, after which the useful behavior of caring for grandchildren evolved in postmenopausal women.

Another wrinkle in a neat and tidy view of evolution is that traits can evolve for one purpose and then change function over time. Sometimes the new function obscures the reason the trait emerged in the first place. A great example is bird feathers. Although feathers are required for flight in birds, most now agree that they did not evolve as an adaptation for flight. In fact, from what we know about the first feathers in the fossil record—from dinosaurs, not birds—they could *not* have supported flight. They may have served as insulation, as courtship signals, or as both—we may never know. Only later, after a series of changes in feather development, could feathers have been used for flight. In later chapters, I'll discuss breasts and orgasms as traits that likely changed functions over evolutionary time, making it difficult to untangle how and why they evolved in the first place.

Traits can also be adaptive in one environment but neutral or even maladaptive in another. When I taught my high school students about peppered moths, I asked them what would happen if we experimentally manipulated the color of those tree trunks. Dark moths are adapted for dark trees, but if we scrubbed those trees clean of soot, dark moths would be at a disadvantage again. In fact, in the real world, the number of dark moths has dropped in regions that have adopted cleaner manufacturing practices, to which they are now mismatched. Mismatch is particularly relevant for human evolution. Our bodies evolved under a certain set of conditions, but recent cultural changes have happened so quickly that our biology hasn't had a chance to catch up.

Our hunter-gatherer ancestors did not have the luxury of going to the supermarket or Super-Duper Burger when they needed to eat. Sometimes food was abundant; sometimes it was scarce. The way the human body metabolizes and stores energy evolved under these feast-or-famine conditions. Our taste for fatty foods and the way we store extra energy as fat may have been adaptations for our ancestors, but today they are liabilities, increasing our risk of obesity and cardiovascular disease. Such mismatches between our evolved biology and modern environments will be mentioned in many of the chapters to follow.

These examples show how the evolution of female biology is more nuanced and complex than the simple example of peppered moth wings. But to appreciate the richness of evolutionary phenomena that were at play in our female ancestors— including the cooperation and conflict I mentioned earlier—we need to return to our high school version of evolution. When I taught my tenth graders about adaptation by natural selection, we discussed predators, food scarcity, and extreme climate or altitude as environmental factors that could drive evolutionary change, but these factors ignore a critical component of our environment—the other individuals with whom we interact. Individuals in social species like ours live in complex groups and must interact with other group members. Even the most solitary mammals interact with other individuals for reproduction, and with their own offspring when they are first born. With these interactions between individuals comes the potential for evolutionary change, because every individual brings their own set of interests to the interaction.

Interactions related to reproduction—the choice of with whom, when, and how frequently—are shaped by sexual selection, a topic that may have been mentioned in your high school text after peppered moths. Sexual selection was Darwin's solution to extravagant traits, like the peacock's tail, which are not easily explained as adaptations for survival. When you imagine a hungry tiger chasing the peacock, with such a conspicuous and cumbersome adornment, it's hard to appreciate how the tail makes it more "fit" to its environment. But if the tail attracts peahens, enabling that peacock to make more babies than the other guy, it may be worth all the trouble of making and living with it. As I'll cover in chapter 5, sexual selection was a powerful force in our evolutionary past, driving the evolution of human body ornaments that include permanent breasts and a large penis, at the same time shaping our perceptions of attractiveness.

In addition to interactions over reproduction, many, if not most, of our interactions are with family. In social species, the groups in which we live consist of family members: parents, children, siblings, aunts, uncles, cousins, grandparents, grandchildren. As anyone who has lived with family knows, family life certainly poses challenges. As one who spends her days dealing with four young children and a harried husband, I can't help but be fascinated by the ways that ancient negotiations and disputes among family members have influenced our biology. As you'll see, the evolution of family relationships in our mammalian, primate, and human ancestors has left lasting traces on our behavior, physiology, reproduction, and aging.

Cooperation is all in the family

The discovery of evolutionary transactions between family members started with honeybees. Honeybees have an unusual social contract. Honeybee colonies contain tens of thousands of bees, but only *one* female—the queen—makes babies, and she is the only one fed royal jelly from the moment she hatches. All the other females in the colony are fed a more diverse diet that leaves them sterile. When the queen is ready, she flies out to find males from other colonies with which to mate. Males don't do much in the colony other than mate with foreign queens—I would say they have a pretty good deal, except for the fact that they die in the act of mating. What do all the other females in the colony do if they're not reproducing? Everything else. When they are young, they keep the hive clean, make royal jelly, and feed all the larvae. When they get older, they act as construction workers, maintaining and expanding the hive when necessary. They become foragers, finding nectar and bringing it back to the hive. As part of the foraging, they become dancers, performing waggle, tremble, and round dances to communicate information about where the food is to other foragers. Worker females also defend the colony, which is why the female (not the male) is equipped with a stinger. The life of a female worker bee is exhausting.

While Darwin was establishing his ideas on natural selection, he puzzled over this social agreement between honeybees, which is known as eusociality. Why would the hard-working female bees leave the job of reproduction to the queen? What's in it for them? Adaptation by natural selection predicts that traits that reduce reproductive success will disappear over time (like the light-colored moth wings during the Industrial Revolution). Sterility is complete reproductive failure; even if you survive longer than anyone, you can't pass on the genes for your longevity to the next generation. So how are sterility and helping behavior maintained in worker bees? Darwin had an answer to this, albeit a somewhat vague one. In his thinking about the puzzle of the sterile worker bee, the concept of *kin selection* was born. He argued in *On the Origin of Species by Means of Natural Selection* that it is possible for traits to evolve that harm you but benefit others if you are all of the same "stock" or family. He hypothesized that natural selection operates on the family as well as the individual.

Kin selection is the foundation for many explanations of social behavior in animals and humans—including the most fundamental of behaviors, parenting—so let me spend a few moments fleshing out the concept and its history. While heredity is a necessary part of natural selection, the importance of genes was not appreciated until Gregor Mendel's now famous experiments on pea plants were recognized in the late

1800s. Scientists Ronald Fisher and J. B. S. Haldane mathematically merged Mendel's genetics with Darwin's selection, and in fact Haldane used the English peppered moths to develop mathematical models of natural selection. In the 1930s, Fisher and Haldane started thinking about the logic of kin selection, and Haldane famously joked that he'd willingly give up his life for two brothers or eight cousins, since on average we share 50 percent of our genes with our siblings and only 12.5 percent with cousins.[1] It wasn't until the 1960s, though, that an evolutionary biologist named William Hamilton presented the full logic of kin selection, which eventually established the theory and popularized it.

I'll briefly describe the math here; the logic is straightforward. Hamilton set it up in terms of costs and benefits, and the currency is reproductive success. The key here is to remember that we share genes with our close relatives; the more closely related you are to a relative, the higher the chance that you both have the same version of a gene. So let's imagine a gene that controls a cooperative behavior that is costly to you but beneficial to a relative (like the costly worker bee behavior of forgoing reproduction to help raise the queen's other offspring). The gene for this cooperative behavior can spread in a population if the cost to you is smaller than the benefit to your relative,

---

[1] This comment was made in an encounter between Haldane and evolutionary biologist John Maynard Smith, who wrote about it in a 1975 article in the journal *New Scientist*. Smith is in fact the biologist who coined the term *kin selection* in 1964.

multiplied by the chance that your relative also carries this hypothetical gene. In other words, if your close relative benefits from your cooperative behavior *and* carries the gene for the cooperative behavior (even if they don't express the behavior themselves), your relative will pass that gene to many offspring. This is how a gene influencing cooperative behavior can spread in a population even if that behavior comes at a cost to the giver.

Critical to the logic of kin selection is thinking about evolution from the point of view of genes, not individuals. All individuals eventually die, but their gene copies live on in their descendants (or in the descendants of relatives who carry the same genes). You may have heard of this as the "gene's eye view" of evolution, popularized by Richard Dawkins in his book *The Selfish Gene*. In the gene's eye view, genes are the players competing for a spot in the next generation. A winning gene might improve the reproductive success of the body in which it's currently being carried or the success of a relative who carries the same gene. Only from the gene's eye view does "selfless" worker bee behavior make sense.

In the 1950s (a decade before Hamilton presented the logic of kin selection), a British biologist named Peter Medawar compared the sterile worker honeybee to a specific group of humans—postmenopausal women. In trying to explain why human females experience menopause, which is a long-standing evolutionary puzzle discussed in chapter 9, Medawar argued that, as with the cooperative genes of sterile worker bees that are propagated through their queen, genes influencing "grandmotherly indulgence" in humans could be propagated through their grandchildren.[2] This was an early outline of the "grandmother hypothesis," which posits that menopause and longer life span evolved in humans because grandmothers who stopped reproducing to help feed and raise their grandchildren were more successful at passing on the genes involved in these traits than women who did not. In addition to the helping behavior of grandmothers, human evolution featured an intensification of social cooperation involving food sharing and a division of labor. As early humans likely lived in small groups of closely related individuals, kin selection is often invoked to explain the evolution of cooperation in our human ancestors. More cooperation in ancestral human groups may have led to the evolution of a longer life span in our species, a trait we'll discuss in later chapters.

While the grandmother hypothesis is still controversial, much evidence from animals (and even plants) supports the idea that cooperative behavior evolves by kin selection. In most cases, the beneficiaries of cooperative behaviors are family members. A study of more than two thousand litters of red squirrels revealed that mothers sometimes adopt orphan pups, but only if the pups are relatives and only if the female's current litter isn't too big, fully consistent with Hamilton's logic. In another example, a dominant male turkey performs an

2   Peter Brian Medawar, *An Unsolved Problem of Biology* (London: H. K. Lewis, 1952), 5n.

elaborate courtship dance to woo females, but instead of going solo, his subordinate brother may help him with the show. The brother never gets the gal himself, but he does help his closely related brother win her over, playing the role of the ultimate wingman. These "selfless" behaviors exist because the genes of cooperative individuals indirectly benefit from behaviors that help close relatives survive and pass on those same genes.

As mentioned earlier, the most basic of behaviors—caring for our children—can be explained by kin selection. Human fathers play a much greater role in parenting their children than do other mammalian or primate fathers. One hypothesis is that paternal care in humans evolved by kin selection (although there are other explanations, as we'll see in later chapters). More universal in primates and mammals is the nurturing and protective behavior of mothers, which exists because of kin selection. We stay up all night to nurse our babies, prepare their food, clean up their messes, drive them to early morning swim practices, mediate their arguments, entertain them, help them with school projects...the list goes on and on. As an exhausted mother of four children, I assure you that these behaviors are costly (in a colloquial sense but also in an evolutionary sense). But we engage in them because they increase the odds that our children will succeed in life and pass down the genes that influence these behaviors to future generations.

By describing motherhood in these genetic terms, I don't mean to reduce all motherly behavior to our genes. The behavior of mothers is complex, flexible, and influenced heavily by our circumstances, something I'll explore in chapter 7. I also don't mean to imply that there is no joy in being a mother. My most cherished moments in life involve my children— nursing and cuddling my newborns, or seeing my son win a swim race after watching him work hard for months at those early morning practices. Truth be told, the joy we feel with our children is likely the result of natural selection itself, since we are more likely to engage in behaviors that our brains tell us feel good. That said, not everything about motherhood feels good. While I took great pleasure in nursing my newborns, I can't say the same about nursing my toddlers, who sometimes bit me and often demanded milk at times I knew they didn't need it. With all my children, there was a point when I had to say enough is enough! And as you'll learn next, there is an evolutionary explanation for why motherly cooperation doesn't last forever.

Conflicts of great interest

Once the logic of kin selection was established, a biologist named Robert Trivers applied Hamilton's theory of kin selection to the special relationship between parents and their kids. But in contrast to Hamilton, who focused on what we're willing to give up for our kin, Trivers focused on the opposite—what we'll fight for. While the traditional view was that parents and children had the same evolutionary interests, Trivers upended that assumption with Hamilton's logic.

Trivers focused on a conflict of interest between parents and their children over parental investment in those children—for instance, the nursing of young mammals after birth, or the feeding of young birds after hatching. In a seminal paper called "Parent-Offspring Conflict," he used the example of a nursing caribou calf to describe this conflict of interest. Like Hamilton, Trivers also looked at the costs and benefits, but his analysis differed in some important ways.

First, the costs and benefits can change over time. When the caribou calf is first born, the benefit to the calf of nursing is huge, and the cost to the mother is small. As that calf grows and requires more milk from the mother, the cost to the mother also grows. Her body becomes depleted, and continuing to nurse also delays the time when she can start on another child, because most mammals do not ovulate while they are nursing.

Second and more importantly, the caribou calf and the mother view the costs and benefits differently. For the mother, the benefit to her current child may come at a cost of fewer children in the future. She's equally related to all her children, so her interest is to invest in a way that maximizes her overall success, perhaps (but not necessarily) by investing equally in all of them. For the calf, though, who *is* the current child, *its own benefit* may come at a cost of fewer future siblings. The calf is 100 percent related to itself but only 50 percent (or less) related to its siblings. So compared to its mother, the calf values the cost of nursing (i.e., having fewer siblings) less than the benefit to itself.

Third, the calf has an active role in its relationship with its mother. The calf is not just passively being nursed—it's soliciting the milk in some way.

What does all this mean for the caribou calf and mother? Early on, when the cost is small and the benefit large, both the calf and mother have similar interests—they both want to nurse. At some point, the mother's body starts wearing down, and as long as she is nursing the current child, she cannot get pregnant again. From her point of view, as her calf grows, a point is reached at which breastfeeding becomes too costly; she's better off weaning her calf and starting on the next one. From the point of view of the calf, it's better off continuing to nurse...but not indefinitely. There is a point further down the line when continuing to breastfeed becomes so costly to the mother that it also becomes too costly for the baby—the benefits to the calf of continuing to nurse don't outweigh the costs of preventing future siblings from passing on the gene(s) the calf might share with them. But for a certain period of time, the mother and calf have an evolutionary conflict of interest over continuing to nurse. That conflict is greater in species in which children are fathered by different males. The less related a child is to its potential siblings, the less it matters if it wears out its mother or delays her production of a sibling, since it's less likely to have those genes passed on through future siblings.

So in the period of conflict between the caribou mother and calf, how does the calf get what it wants? The mother has a clear physical advantage since she is bigger and stronger. What can the calf do? Cry! Newborn mammals cry out of necessity to signal to their parents that they are hungry or cold, but that behavior can also be used in the older child to manipulate mom. As Trivers describes, offspring can use psychological manipulation in these periods of conflict. Anyone who's parented a toddler knows exactly what Trivers is talking about. If I refuse my older toddler a third packet of fruit gummies— obviously something he can survive without—he cries like I'm starving him.

I want to be very clear here about the players in this conflict. I've been referring to a conflict between parents and offspring—the caribou calf tries to manipulate its mother into giving it more milk, or my son tries to manipulate me into giving him more gummies—but we're really talking about a conflict of *genes*. Returning to our gene's eye view of evolution, all individuals eventually die, and all mothers were once children themselves, so we must think about the conflict from the perspective of genes. The effect of natural selection is a change in the frequency of genes over evolutionary time, and the changes in genes expressed in mothers may be opposed by the changes in children. The genetic conflict is what drives the psychological conflict. According to Trivers's logic, genes in babies that promote manipulative behavior during weaning will fare better than those that don't; likewise, genes in mothers that promote timely weaning will fare better than genes that don't.

Based on Trivers's insightful work, biologist David Haig

took parent-offspring conflict even further. In placental mammals, the relationship between mother and baby starts well before birth. Haig describes the conflict that occurs between mother and fetus during pregnancy, again over the amount of resources the mother should provide the fetus. The logic is just what I described for the caribou calf. Both mother and fetus benefit from the fetus getting some amount of resources from the mother—with those resources, the fetus is more likely to survive and pass on its genes, and the mother is more likely to pass on her genes through her fetus. But the evolutionary interests of mother and fetus are not identical.

Again, this happens because a mother is equally related to all her potential children, but a fetus is 100 percent related to itself and only 50 or 25 percent related to future siblings. This sets up the conflict. Fetal genes involved in extracting more nutrients from mothers will fare better than those that take less, because better-fed babies are more likely to survive and pass on those genes. Just as the dark version of the wing color gene spread in the peppered moth population, the version of fetal genes that solicits more nutrients from mothers will spread in a population. But since a mother is equally related to all her children, it is not in her best interest to invest too much in the child she's pregnant with now at the expense of her health and future children, so maternal genes that limit excessive resource transfer will fare better than those that don't.

One difference between the conflict during pregnancy and the conflict after birth is that the fetus has better weapons at its disposal. Instead of using psychological manipulation, which may or may not work (I've become an expert at ignoring my "starving" toddler), the fetus uses chemicals to manipulate its mother directly. During pregnancy, the fetus has direct access to the maternal blood supply into which it can easily deliver manipulative chemicals. As we'll discuss in chapter 6 on pregnancy, there is good evidence that the fetus produces hormones and other molecules that manipulate the pregnant mother into releasing more resources to her fetus than are in her best evolutionary interests to give.

Lest you think that our ancestral mothers were evolutionary pushovers, there is plenty of evidence that they struck back, so much so that the conflict between mother and fetus in the human lineage can be described as an evolutionary arms race. Yes, the fetus produces hormones and other molecules that manipulate the pregnant mother, but the mother has evolved ways to ignore those signals or dampen her response to them. This, in turn, has prompted the fetus to make those hormones in even higher amounts. The arms race of human pregnancy has left us with a puzzle: obstetric complications that can be dangerous for both mother and child, such as gestational diabetes and preeclampsia (high blood pressure).

As you'll learn in many chapters of this book, genetic conflicts between mothers and their children—during pregnancy, childhood, and even into adulthood—have been a major driving force behind the evolution of female biology. These conflicts of interest explain why our pregnancies can be so difficult, why we have periods, and maybe even why we experience menopause. Studying these female traits without understanding the social context in which they evolved can leave us puzzled or misguided about why they exist. Also part of the female story are other genetic conflicts I'll discuss in this book—like that between males and females over reproduction—and of course evolutionary cooperation. Indeed, one factor that has a major impact on our health and well-being in the modern world, our family relationships, also had a profound influence on the evolutionary trajectory of the human lineage, shaping female behavior, physiology, reproduction, and aging. Understanding the evolutionary history of family relationships gives us a deeper knowledge and appreciation of modern female biology and behavior.

CHAPTER 1

The X Factor

During the early days of my fourth pregnancy, I was obsessively preoccupied with having a daughter. With three sons, I had an unending list of girl names desperately waiting for a taker. When sharing with family and friends how much I hoped for a daughter, the responses I received ranged from well-intentioned ("What really matters is having a healthy baby") to questionable ("Boys love their mamas the best!") to downright offensive. The most colorful comment was from the father of a teenage girl: "If you have another son, you will have dodged a bullet!" Apparently, according to this sage dad, teenage daughters are so challenging to parent that it's better not to have them. No matter the wisdom imparted, I held firm in my desire. I yearned for a daughter, a female ally in a home full of males, a daughter to dress in sweet baby dresses, a daughter who would confide in me when she started her periods, a daughter who would come to me first with the news that she was pregnant with her own child.

Although I desperately wished for a daughter in those early days of my pregnancy, I also knew that the "decision" had already been made. In biology, sex is defined by whether an individual makes eggs (female) or sperm (male), and in humans and other mammals, biological sex is decided at conception by which sex chromosome is contained in the father's sperm that fertilizes the mother's egg. With rare exceptions, if the sperm contains an X, the individual develops to be female, and if a Y, male.

The XY system is just one of several ways that sex can be determined by genetics in the animal world. In birds, for example, it's the female egg—with a Z or W chromosome— that decides the sex of the baby. And genetics is not the only way that sex can be determined. In American alligators, the temperature at which fertilized eggs are incubated is what seals the deal—cold makes female, warm makes male. However, for all the creative ways that species might arrive at female and male, the end result is the same: bodies that make big sex cells (eggs) or small ones (sperm)(or in the case of hermaphrodites, bodies that make both—a topic I'll return to later).

While gender is a fluid trait in humans, influenced by cultural and biological factors, the options for biological sex are relatively limited. Have you ever wondered why there aren't any animals with five sexes, ten, or even more? As you will learn, it all comes down to why sex cells can only come in two sizes, big and small, and nothing in between. To better understand why and how our own bodies become female or male, we now take a look at the fascinating evolutionary history of the sexes. Why are there sexes in the first place? Have they always existed? How do our X and Y chromosomes fit into the story of the sexes?

A note on language before our journey

To understand the evolution of the sexes, I'd like to start with some definitions to avoid any confusion on language. In the animal world, at the level of sex cells, there are at most two sexes—no animal (or plant) species produces a third type of sex cell. At the level of the whole organism, other traits such as external genitalia are often used to distinguish between the sexes in animals, but this approach can be misleading—for instance, female hyenas and bark lice have a penis. Humans also develop to make either eggs or sperm. The traits *associated* with biological sex in our species, such as percentage body fat, genitalia, and hormone levels, vary greatly among individuals, and there is some diversity in the combination of traits typically associated with being female or male. As an example, individuals with a rare condition called androgen insensitivity have XY sex chromosomes and testes and produce testosterone, but their body tissues do not respond to testosterone, so they develop female external genitalia. Some individuals with a mix of female and male traits (such as those with androgen insensitivity) refer to themselves as "intersex."

The term *intersex* should not be confused with *hermaph-rodite*, which in biology refers to an individual that makes eggs and sperm in the same body, either at the same time or sequentially in species in which individuals change sex at some point in their life. Hermaphroditism is an evolved reproductive strategy in many plant, invertebrate, and fish species. In some species, everyone is a hermaphrodite, whereas in others, there is a combination of hermaphrodites with males and/or females. No bird or mammal species has true hermaphrodites. In humans, there are very rare cases of individuals who make both ovarian and testicular tissue due to atypical events during early development, but they are unable to make both eggs and sperm. It is not physiologically possible for a human being to be a hermaphrodite, and according to the group interACT: Advocates for Intersex Youth, the term is stigmatizing and misleading.

In contrast to biological sex, gender is a social identity that exists only in humans. It describes how we view ourselves and interact with others in our societies, which is shaped by how our societies define gender roles, relationships, and positional power. While gender is influenced by biology (of the brain, in particular), gender is, by definition, a sociocultural construct. Therefore, gender is potentially a more fluid character than sex.

A world without sex

While I wouldn't describe myself as a science fiction buff, I've watched and read enough sci-fi to notice that the inspiration for many alien species comes from the animal kingdom on earth. The buggers in *Ender's Game* are insect-like queens or workers or drones. Fritz Leiber's *The Wanderer* features large catlike aliens. Many of the beings in *Star Wars* are clearly inspired by animals on earth. The Jar Jar Binks character resembles a dinosaur or bird, and Ewoks are based on a dog breed called Griffon Bruxellois.

Along similar lines, alien species usually have two biological sexes, reflecting the state of biological affairs on earth. Most beings in the *Star Wars* galaxy are female or male. *Avatar's* blue-skinned Na'vi are female or male. While there are deviations from the standard two-sex model in science fiction, many of these are likely also inspired by examples found on earth. The humanlike beings in the classic sci-fi novel *The Left Hand of Darkness* spend three weeks of every month without a sex, but in the fourth week, they can randomly produce eggs or sperm. This brings to mind many hermaphrodite species on earth in which individuals change sex at some point in their lives. One example is the clown fish, that staple organism of all biology textbooks (and *Finding Nemo*) usually photographed in their anemone home. In the short story "Consider Her Ways," women and men existed in the past but a deadly virus wipes out all males, so women learn to reproduce without men by parthenogenesis. Parthenogenesis, a reproductive strategy in which eggs can develop normally without being fertilized by sperm, is used by some earthbound species, including the Komodo dragon. (We'll talk more about parthenogenesis and hermaphroditism later.)

The few sci-fi plots involving an alien species with many biological sexes are true stretches of the imagination, for such organisms do not exist on earth. C. S. Lewis makes a fleeting mention of seven sexes in *That Hideous Strength*, but he didn't describe or expand on these sexes, likely because he thought it would be difficult to convince his readers of a new sex. In a letter to a friend written the year after the book was published, he said, "Try to imagine a new primary color, a third sex, a fourth dimension, or even a monster that does not consist of existing animals stuck together. Nothing happens."[1]

I won't venture to comment on why Lewis or other writers haven't tried harder to develop additional sexes in their science fiction. But I can explain why, at the level of sex cells, we find at most two sexes per species here on earth. To do this, I need to provide some evolutionary context, so we're going to journey back in time to the earth of over three *billion* years ago, a time long before sexual reproduction—or distinct sexes—existed.

This earth looked wildly different than it does today. If you could travel over the land of this ancient earth, you would observe

---

1   In C. S. Lewis, *Letters of C. S. Lewis*, ed. W. H. Lewis and Walter Hooper (London: Collins, 1988), 371.

a barren landscape devoid of anything green or alive. If you could dive into the oceans of this ancient earth, you would similarly encounter nothing. No fish, no coral, no sharks, no seaweed. You would need a microscope to see the one kind of organism then alive—bacteria.

The bacterial species on earth today are incredibly diverse and live in most habitats on the planet—including in our guts and on our skin, in soil, in acidic hot springs, and even in radioactive waste. The bacteria of the ancient earth were also diverse. Some fed on carbon molecules in the oceans, some fed on the energy stored in sulfur molecules, and some used a simple method for converting the sun's energy into fuel for their single-celled bodies. The most important for future life were cyanobacteria, a kind of bacteria still around today that utilize a more efficient process—photosynthesis—to generate usable energy from sunlight. They changed our planet, which initially had little oxygen in the atmosphere (you would have needed an oxygen tank to breathe). In harnessing sunlight to convert water and carbon dioxide into fuel, these cyanobacteria spewed out oxygen as a by-product, forever altering the gaseous composition of the atmosphere and the evolution of life on earth.

How did these single-celled bacteria reproduce and dominate the earth for billions of years? The answer is much the same way that bacteria reproduce today—by cloning themselves. In this type of *asexual* reproduction, or reproduction without a partner, an individual enlarges, makes a copy of its genetic information, which is stored in a circle, and then simply cuts its single-celled body in half. When dividing, it puts one circular DNA copy into each half, resulting in two daughter cells that are genetically identical to each other.[2]

When you think about reproduction in our species, which requires the selection of a partner, successful fertilization and implantation, a *long* incubation, and a painful delivery, cloning yourself sounds like a much easier way to go. But most animals, plants, and fungi do it the hard way—with sexual reproduction. In sexual organisms, babies are made by two parents combining their genetic material. A critical evolutionary steppingstone to sexual reproduction was the evolution of a new kind of cell in the ancestor of animals, plants, and fungi.[3] Rather than letting all parts of the cell float around together in one pool (as in bacteria), the cell of this ancestor had compartments that separated the individual parts of the cell from each other with membranes. Also, its DNA came in linear form instead of circular form. Our cells today still have this design.

---

2   As an etymological side note, terms like *daughter* and *sister* are widely used by biologists when discussing cell division. Bacteria are not female in the strict biological sense of the term—making eggs as opposed to sperm. But because females in the animal kingdom often do the lion's share of the work of reproduction, asexual organisms are referred to as female because they do all the work of reproduction. Also, bacterial cells are more like eggs than sperm in that they provide all the necessary cellular ingredients needed for new life, in contrast to sperm, which only contribute DNA.

3   To be precise, this ancestor gave rise to animals, plants, fungi, and two additional kingdoms that include organisms like algae and amoebas.

The new cell design was important in the transition from asexual to sexual reproduction because it required a new type of cell division to evolve, one you might remember from your high-school biology class—mitosis. Mitosis, just like bacterial cell division, results in two identical daughter cells, but it takes longer and is more complicated, given all the membranes and other special features of these cells. Many single-celled organisms today reproduce asexually by mitosis, including some species of algae. And your entire body, except for your sperm or eggs, was produced by mitosis, including millions of cells in your bones, skin, lungs, and gut that are duplicating right now as you read these words.

Primordial sex

Mitosis was, and still is, a key form of asexual reproduction. But in some of these membrane-containing, mitotically reproducing creatures of the past, something happened that changed life forever. Meiosis evolved. It may not sound like much—it may even recall mind-numbing lectures in biology class—but the evolution of meiosis was huge. Without meiosis, sexual reproduction would not have evolved, and without sexual reproduction, the world would be a completely different place.

You can think of meiosis as a duplicated and modified mitosis. Instead of one cell division, there are two, which halve the genetic information in these cells. In our own bodies, meiosis makes our eggs and sperm. During fertilization, egg and sperm come together, restoring the full amount of genetic information delivered to the offspring. If we didn't halve the genetic content of our sex cells, every generation would contain twice as much DNA as the prior one. You don't need to be a mathematician to deduce that this DNA doubling would quickly spiral out of control!

Meiosis is thus the genetic essence of sexual reproduction, which combines the genetic information from two different individuals. The question of why sexual reproduction evolved in the first place is one of the central problems in evolutionary biology. The costs of sexual reproduction are extremely high: You pass on only half your DNA to your children instead of all of it, and reproduction takes longer and is much riskier. So it is puzzling why it emerged in the first place. But there are certainly evolutionary advantages to sexual reproduction. It doesn't just bring together genetic information from two different individuals; it also mixes the information together in novel ways. This mixing, or recombination, happens during meiosis when matching chromosomes pair up and exchange bits of DNA. The result is that every offspring produced by meiosis is genetically unique.[4] The genetic variation that results from recombination is thought to be beneficial to organisms, allowing them to better respond to changes in the environment. For example, in lab experiments on an algal species that

---

4   The exceptions, such as identical twins, result from the mitosis of a single fertilized egg early in development.

can reproduce sexually or asexually, the sexual populations grew much better on novel food sources than the asexual ones.

I won't plunge more into the debate on sexual reproduction, except to say that it is so prevalent in the tree of life that without a doubt, it is a winning evolutionary strategy. Here the important point is that the evolutionary conflicts I mentioned in the introduction—between mothers and children over nursing and between females and males over reproduction— exist because the individuals that interact in these relationships are not genetically identical due to sexual reproduction.

While meiosis is a key part of sexual reproduction, distinct female and male sexes did not exist when it first evolved over a billion years ago. The individuals in these early sexual populations produced sex cells that were all the same size and shape—no eggs or sperm. To get a better picture of what our early sexual ancestor was like, I'm going to talk about an organism you might be familiar with if you've ever baked a loaf of bread or brewed a kettle of beer—yeast.

Humans have been using yeast for millennia in our culinary endeavors. Hieroglyphics and archaeological evidence suggest that ancient Egyptians were using yeast to make alcohol and leaven bread over five thousand years ago. Residue on an ancient piece of pottery from Iran indicates that wine making goes back at least seven thousand years. While ancient brewers, vintners, and bakers did not understand what ingredient was transforming fruit and grains into alcoholic beverages and bread, we now know that the fungus *Saccharomyces cerevisiae*

is the hero. Fungi like *S. cerevisiae* need food to live and reproduce, just as we do, and they harvest their energy from many of
the same foods that we love, including fruit and grains. Unlike
humans, however, fungi can survive without oxygen. Scientists
have worked out that when *S. cerevisiae* lack oxygen, they are
less efficient at harvesting energy from their food, and they
generate carbon dioxide and alcohol in a process called fermen-
tation. The carbon dioxide helps bread rise during baking, and
the alcohol can be consumed directly, with additional flavors
imparted by the specific yeast strain and sugar source.

While our ancient sexual ancestor was certainly not a
fermenting yeast, we think this ancestor reproduced much
like yeast cells do now. Yeast cells usually reproduce asexually by mitosis, making two daughter cells that are identical to
the parent cell. But under stressful conditions, like starvation,
they switch to meiosis. This supports the point made above—
sexual reproduction generates genetic diversity, which is
advantageous during times of hardship, yielding more genetic
combinations to deal with various types of risk. When a yeast
cell undergoes meiosis, it makes sex cells that contain half the
genetic information of the original cell, just as eggs and sperm
contain half the number of chromosomes as all the other cells
in our own bodies. In contrast to eggs and sperm, all yeast sex
cells are the same size and shape—but they are not identical.

A closer look reveals two types of yeast sex cells, termed *a*
and *α*. The main difference between *a* and *α* is that they produce
different pheromones, or yeast perfume. The perfume each

makes only attracts the *other* cell type. When an *a* cell smells the perfume from an *α* cell, or vice versa, the two cells project toward each other and fuse (copulating yeast cells have been given the least sexy name possible—shmoos). Yeast biologists refer to *a* cells and *α* cells as mating types, and sexual reproduction can only happen between *different* mating types. So in yeast cells, at least, opposites really do attract. Many fungi have only two mating types, but some mushrooms have more than twenty-three thousand. Science-fiction writers should pay heed!

What determines whether a yeast cell will be an *a* cell or an *α* cell? Genes. In *S. cerevisiae*, a set of genes directs the development of mating types, which is similar to regions of the X and Y chromosomes that determine sex in mammals. And if these mating types are starting to sound like female and male, they should. Mating types are not only the yeast equivalent of sexes, but they were likely a critical evolutionary stepping-stone to distinct sexes. As you'll see next, only minor tweaks of the mating type genes are required to start making sex cells of different sizes, and the evolution of big sex cells (eggs) and small ones (sperm) set the stage for the evolution of the sexes as we think of them today.

All sex cells great and small

Before continuing on our journey of the evolution of the sexes, I want to pause to talk about the huge ramifications that sexual reproduction, in particular the evolution of eggs and sperm, had on the planet. Much of the beauty in our world resulted from it—the shimmery ruby-red throat of a broad-tailed humming-bird, the ornate plumage of a peacock, the exotic spectacle of a purple passionflower, the majestic antlers of a male deer. On a practical level, without sexual reproduction, there would be no omelet for breakfast, milk with your cereal, or apples and peanut butter for a snack. Most obviously, if sexual reproduction hadn't evolved, there would be no sex! There would be no young love, no bonding with a lover or spouse, no sexual desire or sexual pleasure.

So how do we get from yeast shmoos to the Kama Sutra? The pivotal event that drove the evolution of the sexes was the origin of big and small sex cells from those that were all the same size. This happened many, many times in different parts of the tree of life. Some individuals in these yeast-like populations in the past started making sex cells that were a little bigger than those of the others. Bigger sex cells possibly made more of the perfume that attracts partners, or they were better able to provision the baby cell to come. As big sex cells started to evolve, so did small ones, which can swim more easily to the source of the perfume. You can make many small sex cells with the same amount of energy that it takes to make one big one, and they can swim farther to find more partners to shmoo with.

At the genetic level, it doesn't take much to go from yeast-like mating types to sex cells great and small. Green algae neatly display this transition because two closely related species exist, one with equal-sized sex cells and one with eggs and sperm. Both species have a similar set of mating type genes that they inherited from a common ancestor. The gene set in the species with eggs and sperm is slightly expanded, but the key difference is in just one gene that both species have. In the species with equal-sized sex cells, the gene prevents one mating type from turning into the other. In the species with eggs and sperm, this same gene acts as a switch, controlling the development of sperm instead of eggs. In some clever genetic engineering experiments, deleting this switch gene in males resulted in egg production, and putting it into females resulted in sperm production.

The specific route to eggs and sperm in green algae is only one of many taken in sexual organisms. The fact that eggs and sperm have evolved repeatedly shows that it's not too difficult to do mechanistically. Importantly, though, the evolution of eggs and sperm did not always result in two separate sexes—it often gave rise to hermaphrodites. Hermaphrodites aren't just the stuff of Greek mythology. The strategy of making both eggs and sperm in the same individual is quite common, occurring in many plants and even in some animals, like the clown fish mentioned earlier. In fact, evidence indicates that when eggs and sperm first evolve in a population, the individuals start out as hermaphrodites. You can think of the brewer's yeast that we've been discussing as a quasi-hermaphrodite, making both *a* cells and *α* cells in every meiosis. One big advantage of being a hermaphrodite is that you have choices: If there are other individuals around and they're easy to find, you can reproduce with them; if not, you can do it with yourself. For organisms that can't move around to find mates (like plants and corals), hermaphroditism is an ideal reproductive strategy. But separate sexes did evolve from self-fertilizing hermaphrodites multiple times—in some plants and most animals—so there is clearly a cost to doing it with yourself. For the same reason that humans avoid having children with close relatives, inbreeding in hermaphrodites can result in low-quality offspring. In many lineages, usually in the complex, multicellular organisms where there are more opportunities for something to go wrong during development, the evolution of separate sexes was favored.

When a species first commits to making separate sexes, females and males resemble each other. But over time, differences accumulate both in the equipment used to make the sex cells *and* across the rest of their bodies. We see this in the green algae mentioned above, which have accumulated additional genes that are active in a sex-specific way. Many body differences between the sexes, such as overall size, body ornaments, and behavior, can all be traced back to the evolution of eggs and sperm. Why? Because eggs are expensive and sperm is cheap.

Let's unpack this statement by thinking about chicken eggs for a moment. Bird eggs contain so much energy that our ancestors were eating them for breakfast long before our species even existed. A hen that lays an egg invested heavily into making it, providing enough energy to her future baby chick for it to survive and pass on her genes. While a productive hen can lay hundreds of eggs a year at an egg farm, this number pales in comparison to the *billions* of sperm that a rooster delivers in *one ejaculation*. Since sperm are so cheap to make relative to eggs, roosters have a different mating strategy than hens (especially in the wild, where humans haven't been artificially selecting for traits we want in our chickens). A rooster is best off sharing his seed with as many hens as possible, which may involve engaging in elaborate shenanigans to impress them or fighting off other interested roosters. A hen, on the other hand, is expected to be picky about her choice of rooster. She will choose a male whom she senses will sire chicks that have the best chance of surviving and reproducing themselves. If she chooses a dud, she will have wasted all that energy! She might prefer the male who impresses her most during courtship (roosters dance in a circle with one wing down), or the one who has the largest and brightest red comb, or perhaps the one who doesn't get sick.

Bird eggs are especially large and energy-rich compared to those of other vertebrates, but the logic holds for all sexually reproducing species that make eggs and sperm, including our own. The difference in the cost of eggs versus sperm is the ultimate reason for many physical and behavioral differences between females and males in any species. This includes both the differences in the machinery used to make eggs and sperm and the secondary characteristics that are associated with being female or male. As we'll discuss in future chapters, many of these secondary sex differences are the outcome of sexual selection, the type of natural selection that is driven by mating success. The flashy bright red comb and flamboyant one-winged dance of that one male rooster are so irresistible that he gets to cavort with many hens. Because of heredity, those hens will have sons who also have flashy bright red combs and dance flamboyantly, and they will go on to attract the hens of the next generation. These kinds of sex-specific physical and behavioral traits—the male comb and dance, *and* the female preference for those male traits—result from sexual selection, which was set in motion the moment that first difference emerged between female and male: the size of their sex cells. Big and small sex cells catalyzed the differences we now observe between female and male.

Back to the big question posed at the beginning of the chapter: Why are there two biological sexes? It all comes down to why there can only be big and small sex cells and nothing in between. Sex cells of intermediate size never make it in the selection—something bigger always does a better job of attracting good sperm and/or provisioning the baby, and something smaller/more numerous always does a better job of swimming to more eggs. In the many times that eggs and sperm have evolved, the end result has always been the same. Two sex cells, great and small.

XY and other tales

We've been wandering in deep evolutionary time, considering events that happened hundreds of millions to billions of years ago. Let's return to the present for a moment and ask a (relatively) simple question. How does a human, like my fourth child, develop into a female or male?

We all know the basic recipe—XX makes female, XY makes male. You get an X chromosome from your mother and either an X or a Y from your father. But our sex chromosomes don't do much in the first couple months of embryonic life. We begin life with the potential to become female or male. The action begins in the tissue that will become our ovaries or testes, known as gonads. Before the sixth week, both female *and* male gene programs are simultaneously active in the gonads. The early embryo also makes female *and* male sex ducts in preparation for either fate.

On the Y chromosome is a gene called *SRY* (which stands for sex determination region Y). It's also known as the master switch, for reasons that will soon become clear. Expression of the master switch gene in the baby gonad tips the balance toward a male fate. If you inherited a Y chromosome, the switch turns on in the sixth week and the male network starts to take over, amplifying the bias toward testes and repressing development of ovaries. Without expression of that switch, the balance tips the other way, toward ovaries.

I don't want to give the impression that making ovaries is the passive option. The female gene program actively promotes development of ovaries and actively represses development of testes. There are a handful of genes that direct ovary development, and if they are missing or not working properly, ovaries do not develop. One gene on the X chromosome is the anti-testes gene—when it's expressed, testes don't develop. I wish I could call it the "X factor" to have a catchy female counterpart to the "master switch," but a whole set of genes is needed to successfully make ovaries instead of testes. This is true for making testes as well, but since *SRY* is on a chromosome that only males carry, it sits at the very top of the sex-determining cascade, acting as the master switch for sex determination.

Notably, though, *SRY* is just a switch for testes development—once the testes form, the gene turns off. So how does sex develop in other parts of the body, including the sex ducts, genitalia, and the brain?

Anyone who has been through puberty knows how potent sex hormones are, and many of the same hormones active during puberty are active in the developing baby. After testes or ovaries form in the embryo, they start to make sex hormones, which then send feminizing or masculinizing signals to the other developing tissues in the body. In males, the baby testes start to produce testosterone and other hormones, which cause the male sex ducts and external genitalia to complete development and the female sex ducts to regress. In females, estrogen produced by the ovaries is involved in the development of the fallopian tubes, uterus, cervix, vagina, and vulva. It takes weeks for these signals to do their work—the earliest you can observe genital differences in an ultrasound is at twelve to fourteen weeks of pregnancy.

Development of nonreproductive traits like body size, vocal cords, and musculature is also affected by hormones produced in the ovaries and testes, although many don't fully develop until puberty produces another wave of sex hormones. An important organ that undergoes sexual development is the brain, which produces its own hormones as well as nerve signals. A large body of evidence shows that sex hormones influence brain development during critical windows of time. In an iconic study from the 1950s, treatment of pregnant guinea pigs with testosterone resulted in their daughters having the sexual behavior of males as adults. Another study showed that exposure to testosterone in a different window of pregnancy resulted in daughters that couldn't ovulate as adults because their brains had developed male patterns of neurohormone release.

In humans, testosterone "treatment" of females happens naturally in cases of male-female twins, where a sister is exposed to a brother's testosterone in the womb. Many studies have found that females born with a male twin are more likely to be rule breakers and exhibit aggressive behavior as adults. A recent study on over seven hundred thousand individuals from Norway, about fourteen thousand of whom were twins, found that females with male twins were slightly less likely to graduate from high school or college, were slightly less likely to get married, and had slightly lower fertility. After

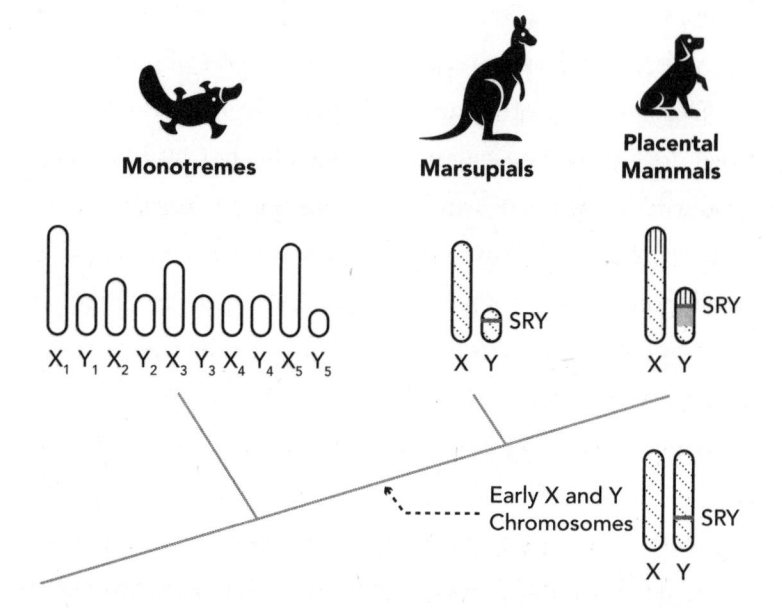

Sex chromosome evolution in mammals. Monotremes like the platypus have ten sex chromosomes (white). Our X and Y chromosomes were born in the ancestor of marsupials and placental mammals from non-sex chromosomes (early X and Y with diagonal hatch marks). They evolved slightly differently in the marsupial and placental lineages (addition of segments with vertical hatch marks in placental mammals). In most species, the Y chromosome has degenerated and is much smaller than the X.

controlling for some of the other factors that could explain these trends (like simply growing up with a brother the same age), the authors point to prenatal testosterone as the culprit. As a female with a twin brother myself, I'm glad that I looked at these data after finishing school and having kids!

Biological factors other than hormones are also thought to influence the sexual development of the brain. There is some evidence in mice and humans that the *SRY* gene is turned on in parts of the developing brain, where it could influence masculinization independent of hormones from the testes. It also appears that the number of X chromosomes you have in your brain cells—one (male) or two (female)— affects development of the brain in ways that aren't well understood (see below). Since our brains influence our sexual identities, affecting traits such as gender and partner preferences, there is considerable interest in understanding the role of hormones, genetics, and factors like culture and experience in the sexual development and function of the brain.

One of the ways this is being studied is in animals that are more flexible about their sexual identities. In some species of hermaphroditic fish, individuals can switch from making eggs to sperm (or vice versa) without changing their female or male behaviors. They can also change their behaviors before switching the type of sex cells they make. This shows that sexual development and function of the brain, at least in these species, are more flexible than making a uterus or prostate gland.

Sexual development of the brain is also being studied in humans with a disorder of sexual development (DSD). DSD is a medical classification that includes over fifty conditions; some of those diagnosed with a DSD prefer the term *difference of sexual development*, and others prefer the term *intersex*. I mentioned at the beginning of the chapter that there are some exceptions to XY sex determination in humans. These exceptions involve DSDs, which can result in a mismatch between anatomical, psychological, gonadal, and/or chromosomal sex. The most obvious DSDs involve extra or fewer sex chromosomes (as in individuals with XXY), which result from errors in meiosis in the mother or father. Brain images of individuals with extra sex chromosomes show that the number of X chromosomes affects the size and shape of different regions of the brain, which may influence sexual identity in ways that are not understood. There are other DSDs caused by missing, extra, or altered genes that play a role in sex determination. In XY individuals, having an extra copy of one of the genes that sit at the top of the ovary-making cascade can result in an ovary-like gonad and a rudimentary uterus. Similarly, if another gene involved in making ovaries isn't working properly in XX individuals, they develop a gonad with both ovarian and testicular regions. I mentioned androgen insensitivity earlier, a DSD that affects everything downstream of testes determination. XY individuals with androgen insensitivity make functional testes, but the rest of their bodies can't properly detect or respond to the hormones produced by the testes. Thus, individuals develop female external genitalia and often consider themselves female, even though they have a Y chromosome and testes and make testosterone.

While DSDs reveal more variation in human sexual development than was previously acknowledged, this variation pales in comparison to the astounding diversity of sex determination mechanisms found in nature (both across and within species). The number of ways that sexual organisms become female and/or male is bewildering. Just looking at fish, which are especially flexible in their sex determination strategies, you find species with XY chromosomes like ourselves (for example, brown trout females have an XX, males an XY). You find species with ZW chromosomes (finless eel females have a ZW, males a ZZ).[5] You find species with *different* XY or ZW chromosomes, which means that their sex chromosomes evolved independently (rainbow trout have a different pair of XY chromosomes than the closely related brown trout; and on a related note, the XY chromosomes of both trout species are different from our XY chromosomes, which had an independent evolutionary origin). You find species that are transitioning between an XY and ZW system and are using both mechanisms at the same time (blue tilapia). You find species that determine sex genetically but have no obvious sex chromosomes. You find species that decide sex by temperature or other environmental factors such as food or nest availability. You find species that have dispensed with males altogether, females reproducing by a strange type of parthenogenesis, in which eggs can develop normally without being fertilized

---

5   The difference between an XY and ZW system is simply who is determining the sex of offspring. In an XY system, males have two different sex chromosomes, so the sex chromosome contained in sperm decides offspring sex. In a ZW system, females have two different sex chromosomes, so the egg is the decision-maker.

by sperm. And you find *many* hermaphrodite species. In the last section, we talked about the evolution of separate sexes from hermaphrodites, but in more recent evolutionary times, the opposite has happened too—separate sexes revert to being hermaphrodites. In some hermaphrodite fish species, individuals start out as male and then change sex to female (like the clown fish mentioned earlier); some start out as female and change sex to male (the bluehead wrasse); some appear to make eggs and sperm at the same time (the blue-banded goby). My favorite hermaphrodite is the chalk bass, a monogamous fish that switches back and forth between male and female multiple times a day! The chalk bass's partner does the reverse, in a spectacular sex-changing lovefest.

And I've only been describing fish. When you zoom out and look at all sexually reproducing organisms, including insects, reptiles, and plants, the variation in sex determination strategies is dizzying. Remember the honeybees that we discussed in the introduction? In bees, ants, and wasps, males develop from unfertilized eggs and females from fertilized eggs, giving males half the genetic content of females and mothers control over the sex of their offspring, which is yet another strategy to decide sex. Clearly, once eggs and sperm evolved, species got creative and flexible about how to make them. Ecological conditions and/or social factors may drive the switch to a different strategy. In a species with females and males where it becomes difficult to find sexual partners, changing to hermaphroditism might be a better evolutionary option. It might even be best to dispense with sex altogether and revert to asexual reproduction. I mentioned the fish species that use parthenogenesis, but there are others too. Some permanently changed to parthenogenesis (several salamanders, geckos, and lizards), while others switch back and forth between sexual and asexual reproduction, depending on the availability of males (mayflies are a well-studied example, but Komodo dragons, aphids, water fleas, and hammerhead sharks sometimes use parthenogenesis for reproduction).

Given how easily and quickly sex determination evolves, it would be very difficult to chart the entire evolutionary history of the sexes from the beginning to our species today. That said, with this astounding diversity as our backdrop, it is notable that sex determination in mammals is so uniform. Mammals are relatively square when it comes to making the sexes— all mammals reproduce sexually, and the vast majority use the same X and Y chromosomes to decide sex. No mammalian species has evolved parthenogenesis, hermaphroditism, a different set of sex chromosomes (although stay tuned on this), or the use of environmental cues to determine sex. In mammals, sexual reproduction—and sex determination by the X and Y—have been locked in.[1]

---

[1] Sexual reproduction has been locked in because of genomic imprinting, a mammal-specific phenomenon that we will discuss more in future chapters. It is not clear why the mammalian X and Y are so stable compared to the X and Y of other groups (as in some fish).

This "squareness" of mammals is a gift to biologists interested in sex chromosome evolution, because it allows us to trace the history of the X and Y chromosomes in this group. By comparing the sex chromosomes of all three major subgroups of mammals—monotremes (egg layers like the platypus), marsupials (kangaroos), and placental mammals (most mammals)[2]—we know that our X and Y evolved from a pair of regular chromosomes in the ancestor of all placental and marsupial mammals. The third and earliest branching group of mammals, the egg-laying monotremes, have a complicated sex determination mechanism involving ten sex chromosomes. At the beginning, our own X and Y were identical in size and content (just like our twenty-two pairs of non-sex chromosomes). But then a series of events changed them to what they are today.

First, the X and Y chromosomes were born with the birth of the master switch gene. Remember, this gene is called *SRY*, and it is only found on the Y chromosome. *SRY* is very similar to another gene called *SOX3* on our X chromosome. Research shows that *SRY* first evolved when a mutation in

---

2   The term *placental mammal* has been used for decades to describe the group of mammals that make a complex invasive placenta. It was recently discovered that marsupial mammals make a simple and short-lived placenta, so the group that was previously referred to as placental mammals is now officially called eutherian mammals. I'm trying to keep the terminology simple here, so I'll stick with the term *placental*, which I use to describe mammals with a complex invasive placenta, not including marsupials.

SOX3 turned it on in the baby testes. There was certainly another sex-determination mechanism in place before this event (in other words, there were females and males), although your guess is as good as mine on what it was.

Second, the Y chromosome started picking up male-specific genes. If you look near the human *SRY* gene, you'll find genes that play a role in making sperm. What often happens during sex chromosome evolution, not just in mammals but in any group with sex chromosomes, is that additional genes start hitchhiking along with the master switch gene. These genes were either picked up from other chromosomes, or they were genes on the X that mutated to benefit males specifically, just like the original master switch mutation.

Next, as these male-beneficial genes started accumulating near the master switch, a key event during meiosis *stopped* happening—recombination. I mentioned before that meiotic recombination is one of the advantages of sexual reproduction, jumbling together bits of the chromosomes you inherited from your parents in novel ways. But as genes on the Y started picking up other male-specific genes, recombination in the area of the hitchhiking couldn't happen because the X and Y chromosomes couldn't pair up properly during meiosis.

Then the Y chromosome started to shrink. The human X chromosome has about a thousand genes, but the stubby Y chromosome only has dozens. This may surprise you, since I just mentioned that the Y added some male-specific genes. So how did it become so small? One of the many benefits of recombination is that it helps eliminate damaging mutations that are always popping up in our genomes. The X chromosome stayed healthy because it continued to pair up properly and recombine in females, but in males, some regions without recombination on the Y became damaged beyond repair and were lost. Apparently, the Y chromosome can't have its cake (adding male-specific genes) and eat it too (keeping genes from the X chromosome).

In humans, there are still small regions of the X and Y that can pair up during meiosis and exchange bits. But the story of sex chromosomes in humans differs in other mammals. Some mammals have lost less of their Y chromosome, and some have lost the entire thing. Mole voles and Japanese spiny rats have independently lost the entire Y chromosome. There are still males in these species, so they must have evolved other ways to determine sex. While we don't yet understand how they did it, these examples do reveal rare cases of sex chromosome turnover in mammals.

Such widespread Y shrinkage has suggested to some that the human Y chromosome could disappear entirely. After all, the Y was completely lost in those spiny rats and mole voles. However, the vast majority of mammals have not lost their Y, and studies have shown that the rate of Y degeneration in primates has slowed considerably. So while the loss of the human Y is a remote possibility, the male sex is certainly not doomed anytime soon. But the possibility does make for great science-fiction material.

And in that spirit, let me return to the identity of my fourth child's sex chromosomes. I very happily got my daughter. But perhaps if I had waited long enough (say 4 to 10 million years), another XY baby would have been impossible!

CHAPTER 2

Baring the Story of Your Breasts

Early on in my science training, I took a human osteology lab as a graduate student at NYU. Human osteology is a class for the very few folks in the world who need to be able to identify human bones in their profession, such as human paleontologists who go on fossil-hunting expeditions in places like Kenya and Indonesia and forensic scientists who ID human remains at crime scenes. I spent hours in the lab memorizing ridges and bumps on the more than two hundred bones and teeth in the human body. During our exams, we were presented with slivers or small pieces of bone and asked to identify exactly where they belong in the human skeleton. To the untrained eye, these fragments might fit anywhere. Our sly professor even included some nonhuman bones on those exams to make sure we were paying attention. If possible, we also had to identify to whom the bone fragments belonged. A tween or a centenarian? Someone healthy or sick? A female or a male?

Soon after I completed that osteology lab, I decided I was more interested in genes than bones as a means to study the biological past, so I've forgotten most of the arcane knowledge I picked up that semester. But one of the tidbits I still remember is that the skeletons of immature females and males look identical—it is impossible to distinguish between them. Indeed, without their accoutrements, it's also difficult to distinguish living girls from boys before they've reached puberty. My mother apparently pierced my ears as a baby so that all would know she had girl/boy twins, not boy/boy twins. With her two X chromosomes, my own daughter was born with ovaries and all the eggs that will ever be available to her (a subject I'll return to in later chapters), but at two years old, with her diaper and clothes on, she was often mistaken for a boy. I admit that many days I dressed her in her brothers' beaten-up hand-me-downs, something I'd never get away with now that she's closer to four. Anecdotes aside, my point is that many of the physical traits associated with each sex don't appear until puberty (the exception is the external genitalia, with which we are of course born).

Puberty is a time of major physical change. In females, the maturation of the ovaries and associated hormonal changes lead to the development of breasts, the accumulation of fat in the hip and thigh area, pubic and underarm hair growth, and the start of monthly periods. Human paleontologists and forensic scientists know well that it is also a time when the pelvis undergoes a marked widening that will allow a woman

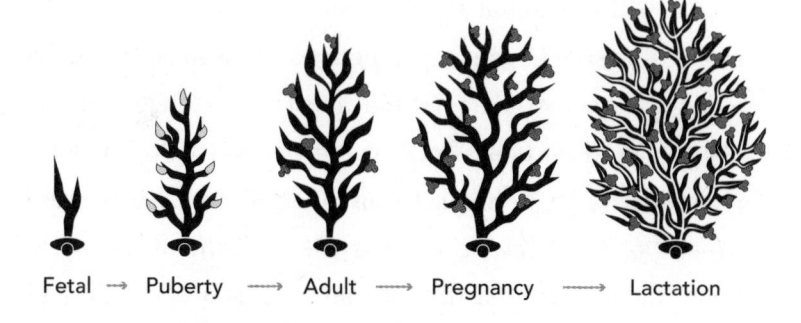

Fetal ⟶ Puberty ⟶ Adult ⟶ Pregnancy ⟶ Lactation

**The blooming of the mammary gland at
different stages of development.**

to give birth—also enabling the lucky paleontologist to infer
a "Lucy" rather than a "Lucius."

I'll discuss periods, pregnancy, and childbirth in later
chapters; here I focus on breasts. After reviewing the key
hormonal and physical changes of puberty, with an emphasis
on breast development, I'm going to tackle two evolutionary
questions of interest. First, why did breasts evolve in the first
place? (The answer is not as obvious as you think.) Second,
why did breasts become a permanent body trait in human
women? In every other species of mammal, including in our
closest primate relatives, breasts are only visible when a female
is pregnant or nursing; they disappear when she is not.[3] In
humans, a woman carries her breasts her whole postpubescent
life, whether or not she ever uses them for nursing a baby.

---

3   The udders of dairy cows are always visible because dairy cows are kept
      in a perpetual state of pregnancy to keep them producing milk.

Shaping up

In the last chapter, we discussed female development in the womb—ovaries first, then the rest of the reproductive tract and the genitalia with the help of ovarian estrogen. Breast tissue also starts to develop, but in contrast to the reproductive tract, prenatal breast development is similar in females and males. After about a month of embryonic life, two parallel ridges form on the belly side of the embryo, called milk lines, that travel from the armpits all the way down to the groin region. Along these milk lines, mammary buds sprout in mammalian embryos. Individuals match the number of buds to the maximum litter size of their species. For example, mice make ten buds (I know from my work with lab mice that they have litters of roughly six to ten pups), and humans only make two. As buds form, the unneeded stretch of each milk line disappears in human embryos. You do occasionally see extra nipples on people, as in beach shots of self-assured actors like Mark Wahlberg, or if you're a fan of the sitcom *Friends*, you'll know about Chandler's "nubbin." These are rare cases in which the superfluous section of milk line didn't completely regress and an extra nipple formed. During the remainder of prenatal development, the foundation of the breast is established beneath each bud—some fat cells, a rudimentary gland that will make milk, ducts that will move milk, and a nipple. Both female and male newborns can actually produce milk, and about 5 percent of babies do! It's been called witch's milk, the name derived from seventeenth century folklore that described witches stealing this milk to use in their magic. Once any lingering maternal hormones pass out of the newborn's system, witch's milk dries up, and breast development pauses until puberty.

After we are born, our bodies and brains grow, but not much happens to the breasts or reproductive system during infancy and early childhood. This is true of most organisms—there is typically not enough energy to grow larger and reproduce at the same time, so organisms do one or the other. Postnatal brain growth is especially notable in our species. Humans have famously large brains for our body size, but bipedalism puts a constraint on how wide the pelvis can be and thus how large our babies' heads can be at birth. A narrow pelvis is better for walking and running on two legs, so the human brain completes much of its development *after* birth. Humans also have a long childhood growth period compared to other mammals. As a species, we invest much time and energy into growing larger bodies and brains, reaping the evolutionary benefits later in life when we are able to live longer and produce more children.

We need to talk about puberty to understand breast development, since the hormonal changes at puberty drive changes in the breast. Girls take their first steps toward sexual maturation earlier than you might think, at about six or seven. It starts with the adrenal glands, which are like little beanies sitting on top of our kidneys and are best known for making adrenaline, hence the name. But they make other hormones too, including the adrenal androgens that play a role in sexual maturation. You can thank the adrenal androgens for some of the less savory aspects of puberty (at least according to Western standards), like body odor, acne, and pubic hair. Testosterone is an example of an androgen, but it's made by the testes. In females, the adrenal glands are the main source of androgens, which are produced in tiny amounts in the six- or seven-year-old girl (it's a little later in boys), gradually ramping up and peaking around age twenty. This happened to be the age I was when, completely fed up with my body's androgen production, I marched into the dermatologist's office, demanding something to clear up my skin.

A few years after the adrenal glands begin to mature, the ovaries resume their development and start pumping out estrogen. Estrogen is the key hormone for breast development during puberty, so let's discuss the events that kick-start its production by the ovaries. There are three hormones made in the brain that prompt ovary maturation. Gonadotropin-releasing hormone (GnRH), which is produced in the hypothalamus, tells the pituitary gland to make and release follicle-stimulating hormone (FSH) and luteinizing hormone (LH), which leave the brain to stimulate the ovaries. The brain and ovaries communicated with these same hormones during prenatal development, but soon after birth, signaling was shut down. So the hormone signaling that initiates puberty just reactivates a system used early in life.

How does the brain know when to start making the

hormones that activate the ovaries during puberty? We don't know precisely, but there are likely many factors involved, including genetics (women often get their first period around the age their mothers did), weight (overweight girls get their first period a little earlier than average, and anorexic girls don't menstruate), general health (healthy girls get their first period earlier than those with underlying health issues), and a girl's stage of skeletal growth (successful reproduction requires that the pelvis is large enough to allow a baby to pass, which is why very young girls who aren't close to completing skeletal growth don't start menstruating). Whatever the ultimate triggers, once the brain starts talking to the ovaries again, the ovaries continue their development. They resume making estrogen, which continues its work maturing the reproductive system and the rest of the body. This doesn't happen overnight. It takes years for a girl's reproductive system and outward physical appearance to mature. Maturation is usually complete by around fifteen to seventeen.

While estrogen does essential work inducing the growth and maturation of the internal reproductive organs (the uterus, vagina, and the ovaries themselves), the noticeable changes of puberty are on the outside. Here is the typical sequence (with much variation in the absolute timing and the order of these events): A girl develops breast buds; she starts growing pubic hair; she has a growth spurt; she gets her first period; she grows some underarm hair; she starts accumulating fat around her hips and thighs; her breasts grow to their adult size. The onset of menstrual periods is often taken as a sign that a girl has reached sexual maturity, and it is possible for a girl to become pregnant once she starts menstruating (and even before she gets her first period), but most menstrual cycles in the first year or two are *anovulatory*—no eggs are released. Only toward the end of this multiyear sequence of outward changes does a female start ovulating regularly and reach sexual maturity.

Estrogen from the ovary is the main signal that initiates the internal and external changes during puberty (but remember, adrenal androgens are also involved, as well as hormones I haven't discussed, like growth hormone and progesterone). Estrogen is made from cholesterol, which has a bad reputation for clogging arteries, but it's an essential molecule from which many kinds of steroid hormones are made. While the brain is the organ that nudges the ovary to make estrogen (via its hormone messengers FSH and LH), the brain senses estrogen as well, deciding when to release FSH and LH based on how much estrogen it's sensing. So the hormone signals that initiate puberty and menstrual cycles throughout life are more like a back-and-forth conversation between the brain and ovaries rather than a lecture by the brain. Even the ovary, which makes estrogen, responds to estrogen too. It is important that many of the tissues in our bodies, including the brain, ovary, breast, and uterus, can sense and respond to estrogen.

Let's now consider how estrogen elicits changes in breast tissue during puberty. Remember that at birth, the basic components of the breast are present in females and males.

During puberty, estrogen transforms this simple breast into the adult breast—a complex matrix of fatty and fibrous tissue, within which is embedded the equipment needed to make and transport milk to a baby.[4] A tree analogy helps to visualize the structure of the milk equipment. You can think of the ducts that carry milk in the breast as the branches and the glandular cells that make milk as flowers at the end of the branches. The treelike network of glands and ducts leads to the nipple, the base of our imagined tree. In this tree, nectar—i.e., milk—flows from the flowers at the tips of the branches down to the base of the tree.

Sticking with our tree analogy, the infant breast is the tiny, bare tree when you first planted it in your yard—very few branches and no flowers. At puberty, the breast starts looking more like a tree. New branches form in response to ovarian estrogen, and the tree starts making flower buds but not flowers yet. The glandular cells proliferate and become able to make milk during puberty, but they need additional hormone signals to bloom, which will come during pregnancy and nursing. However, the visible changes to the breast at puberty have little to do with transformation of the milk equipment. The visible changes result from a transformation of the matrix *around* the ducts and glands.

---

4   Males don't typically make enough estrogen for this transformation to occur, but in rare cases where estrogen levels are high in males, they will develop visible breasts.

In response to estrogen from the ovaries, the matrix fills out by accumulating fat.

The amount of fat in the matrix of the breast is what distinguishes an A cup from a D cup, not the space taken up by the glands and ducts. A woman with small breasts is equipped to nourish her infant with just as much milk as a woman with large breasts. What is critical for milk production is the frequency of nursing, best illustrated in women who successfully breastfeed multiple infants, or in women who double milk output from one breast when the other isn't working properly. Fat is the major contributor to the volume of the breast, and large breasts simply have more fat in them than small ones. As I mentioned, estrogen from the ovaries is the main signal to the breast matrix to lay down fat during puberty. But the ovaries of large-breasted women aren't necessarily making more estrogen. Several studies have looked for a correlation between breast size and circulating estrogen levels, but most have found none. Differences in breast size likely have more to do with how sensitive the fat-producing cells of the breast are to estrogen (and to the other hormones that relay messages to the breast).

Much of this sensitivity is determined by genetics. The idea that genes control breast size is an intuitive one—a woman with a D cup likely has a mother or grandmother or aunt with large breasts. Beyond intuition, though, large genetic studies on thousands of women have been conducted that have looked for a link between differences in our genomes and differences in breast size (and breast cancer risk). These studies have found that genetic differences in and around the estrogen receptor gene, along with a few other genes, are correlated with differences in breast size (and breast cancer risk). The estrogen receptor is a protein in our cells that senses estrogen and acts in response to it. So making more or less of this estrogen sensor in fat cells in the breast, or perhaps tweaking its structure and/or activity, might be one evolutionary route to larger or smaller breasts.

While we have some hints as to how variation in breast size is encoded in the genome, it is unclear why there is such variety in human breast size and, even more fundamental than that, why there is fat in our breasts at all. While many of the changes during puberty, such as widening the pelvis, enlarging the uterus, and growing the network of ducts and glands that will make milk, have an obvious reproductive purpose, it is not clear why our pubescent bodies need to lay down fat in the breasts (and hips and thighs) at all. Humans are the only mammals who do this, which has prompted much speculation on the evolutionary advantages of fatty breasts. We will investigate this question later in the chapter. Before that, though, I want to ask the more obvious question about breasts—why do mammals grow these trees of glands and ducts in our chests in the first place?

Budding the first breasts

The ability to make milk is a defining trait in mammals.

Well, half of them at least. Male mammals do not lactate, with few exceptions.[5] I mentioned in the previous chapter that mammals are a diverse group of animals with three subgroups: the egg-laying monotremes, the pouch-forming marsupials, and placental mammals. Mammals are distinguished from other types of animals such as reptiles and birds by the presence of hair, sweat glands, a few skeletal differences, and mammary glands, which is the trait taxonomist Carl Linnaeus used to name the group in the 1700s. Linnaeus's Mammalia is translated "of the breast" and is derived from the Latin word for breast, *mamma.*

While female mammals are unique in having mammary glands, other animals have evolved a variety of creative ways to feed their wards when they are too young to acquire or digest their own food. In many bird species, females and sometimes males will consume and digest food, regurgitate it, and transfer it directly to the mouths of their hatchlings. In pigeons and flamingos, females and males make a liquid called crop milk, named after the structure in which it is made. The crop, which is found at the base of bird necks and normally stores and moistens food, secretes a protein- and fat-rich liquid in these species that is used to feed hatchlings. Some marine snails make extra, unfertilized eggs that are eaten by their offspring.

---

5   Male lactation has been observed in two species of fruit bats in the wild, although its prevalence and purpose remain unknown, and in some domesticated animals and humans with hormonal disorders or under the influence of exogenous hormones.

The hatchlings of discus fish feed off the skin of their parents, both of which secrete a protein-rich mucus to feed their young. The female African caecilian, a limbless amphibian that lives in Kenya, also feeds her hatchlings from her skin, which is transformed into a nutritious substance that her babies scrape off her body with their teeth (poor mama! I thought we had it bad with the occasional nip by a nursing toddler, but daily skin scrapings seem like a worse fate). The most extreme cases are found in some species of spiders, in which the mother is eaten alive by her hatchlings for nourishment.

Among these impressive strategies that animals use to nourish their young offspring, lactation stands out like no other. Lactation allows adult mammals to specialize on foods that would be impossible for their babies to catch or digest, and it gives mothers more flexibility in how they live and reproduce. Blue whale mothers nurse their babies for about six months, during which time they don't feed themselves and their calves gain almost forty thousand pounds. Some bears can nurse their babies for up to two months in the den without leaving to eat, all because they are able to stock up beforehand and then tap into their fat stores while nursing. I don't like to admit it, but my children really didn't need me to consume that daily cinnamon bun and latte while nursing, since, like the bear and blue whale, I was well stocked before I started (by consuming so many cinnamon buns and lattes while pregnant!).

The milk that the mammary gland produces is also

exquisitely designed to meet the needs of young mammals. All mammalian milk contains some basic ingredients: fat, carbs, and protein. The specific versions of fat (milk fat globules), carbs (lactose and more complex sugars), and protein (caseins and whey) in mammalian milk are unique—they aren't seen anywhere else in nature. Milk also contains hundreds of other ingredients, including a variety of antibodies, antimicrobial factors, hormones, vitamins, minerals, maternal stem cells, and even bacterial cells. Recent studies on human milk have identified over fifteen hundred proteins that help in digestion, immunity, and neurodevelopment. The complex sugars in human milk are prebiotics, which encourage the growth of beneficial gut bacteria, and antibiotics, which prevent the growth of harmful bacteria. One of the amazing things about milk is that its composition can change over the course of a single feed and over the course of the days/months/years that a mammal breastfeeds, depending on what the infant needs at the time. Rhesus monkey mothers can even change the composition of their milk depending on if they are nursing sons or daughters—sons get more protein and fat, whereas daughters get more calcium and a greater volume of milk. Milk composition also varies across species depending on the specific needs of the infant mammal. Hooded seals born on the ice need fat to survive, and not surprisingly, fat content in seal milk is about 60 percent (human milk, in contrast, only contains about 4 percent fat). The milk of the eastern cottontail rabbit is high in both protein and fat, likely because the

and disseminating his ideas, for many of his contemporaries thought it inconceivable that the complex mammary gland could have evolved for milk production by "numerous, successive, slight modifications"[6] as Darwin's theory predicted. In fact, most experts now agree that breasts initially evolved for function(s) other than nutritional support for baby mammals. Only later, after a series of many slight modifications, did mammary glands take on a nutritive role.

What were these other functions? One of the most plausible explanations for the evolution of the mammary gland comes from Olav Oftedal, a scientist at the Smithsonian. Using a combination of paleontology, comparative and developmental biology, and genetics, he argues that there were many intermediate steps in the evolution of the milk-producing mammary gland. They started over 350 million years ago in the first four-legged animals to set foot on land. These amphibian-like creatures had simple skin glands that secreted a fluid that kept their skin moist and free from infection. Skin secretions may have also kept the eggs of these ancient creatures moist and microbe-free, as we see in modern amphibians today (amphibians secrete hundreds of antimicrobial compounds from their skin glands, and salamanders that breed on land keep close to their eggs to keep them hydrated).

---

6   Charles Darwin, *On the Origin of Species by Means of Natural Selection* (London : John Murray, 1859), 189.

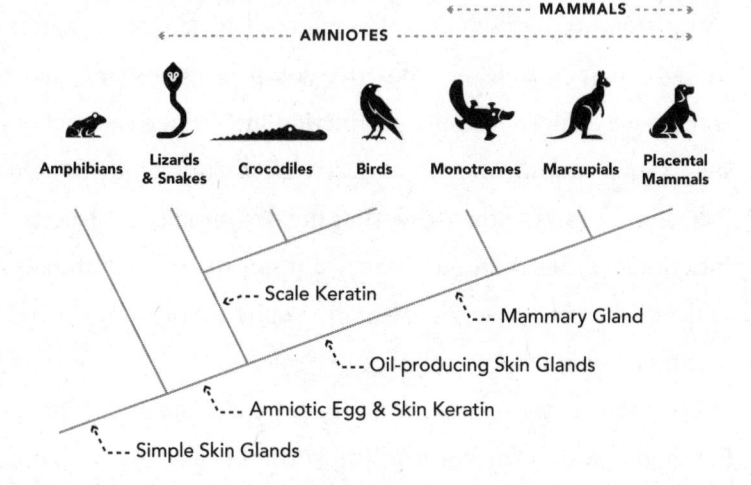

The emergence of key traits in land animals.

Between 300 and 350 million years ago, the first amniotes evolved, a group that includes modern-day mammals, reptiles, and birds. The key distinction between amniotes and amphibians is the amniotic egg, which contains extra membranes that allowed the first amniotes to live and reproduce away from the water. Every time you crack open a chicken egg, you catch a glimpse of this monumental evolutionary development. The fertilized amniotic egg has a membrane that stores the yolk, a membrane that stores waste, a membrane that surrounds the embryo (the eggs you buy at the store aren't fertilized, so don't worry, they don't contain embryos), and a membrane that surrounds all these compartments and permits gas exchange with the environment. Many amniote eggs, like those of chickens, have a shell, but the eggs of these early amniotes were not covered in a hard, calcified shell like that of chickens. Rather, they had a leathery, parchment-like covering, similar to the covering around a platypus egg. Early amniotes soon divided into two lineages—the line leading to mammals and the line leading to birds and reptiles. Some on the bird/reptile line evolved a hard, calcified eggshell, but the line to mammals stuck with parchment. The problem with parchment is that it allows eggs to lose (and gain) water easily. This makes them susceptible to drying out, so the female mothers of these parchment-covered eggs would have needed to find a solution to this problem.

We know that the skin of these ancient amniotes also had to adapt—dehydrated eggs weren't their only problem. Because they were also losing water through their skin, early amniotes added keratin to their skin, a tough, fibrous protein that acted like a protectant and sealant (this is why keratin treatments for hair are so popular). Birds and reptiles evolved additional, even tougher keratins to make scales, which do an excellent job keeping moisture inside the body. But this wasn't in the cards for our branch of amniotes—we lack the additional, specialized keratin molecules needed to make scales. Instead, new types of skin glands evolved, probably from the simple skin glands of the amphibian-like animal I mentioned earlier. In addition to watery secretions, these new glands secreted substances that helped waterproof the skin. Oily or waxy substances would be suitable for this.

Oftedal calls out one gland in particular—the apocrine sweat gland. We have many apocrine sweat glands on our bodies—under our armpits, in the groin area, and on our heads. They produce an oily sweat (as opposed to eccrine sweat glands, which produce a dilute, watery sweat). Apocrine sweat glands are often associated with one hair and one sebaceous gland (which produces an oilier substance, sebum, which contributes to acne in some folks). Apocrine glands make some of their secretions in a distinctive way—secreting sweat in little fatty sacs that bud off the cells of the gland. This resembles the way the mammary gland secretes milk fats. The basic structures of the apocrine and mammary glands are also similar (a layer of secreting cells underlain by a layer of muscle-like cells that contract to help the gland unload its goods). Given these similarities, Oftedal suggests that an apocrine-like sweat gland was the precursor to the mammary gland. Indeed, the mammary gland of the most "primitive"[7] mammal around today, the platypus, is still associated with a hair and a sebaceous gland, supporting his idea.

At first, these sweat and oil glands secreted substances that lubricated and waterproofed the skin and eggs of premammalian amniotes, and the secretions may have also contained some antibacterial ingredients to keep skin and eggs free from infection. The secretions may have also helped in thermoregulation—we know that ancient amniotes on the line to mammals gradually became warm-blooded and would have evolved mechanisms like

---

7   The platypus is not any less evolved than we are, but the line to the monotremes branched away from the rest of the mammals first, and we think many of their features, like egg laying, were also exhibited by the first mammals.

sweat to help maintain a constant body temperature. While this all makes sense given what we know about amniote evolution, how do we get from oily sweat to milk? Modern genetics helps us understand what happened next.

As with milk, one of the primary jobs of the amniotic egg is to nourish developing babies, and we know some of the genes involved. Let's think about the chicken egg again. It needs to provide food to the developing embryo, most of which is packed into the yolk. Yolk contains proteins, fat, and even some carbs, as well as essential minerals like phosphorus and calcium (just like milk). One of the most important proteins in the yolk is called vitellogenin. Made in the liver of mother hens, vitellogenin is an important source of protein, and it carries many of the other nutrients in yolk. Scientists have scanned the genomes of living amniotes and have found three vitellogenin genes. In reptiles and birds, whose developing offspring still rely on egg yolk, these genes are alive and well. In mammals, though, the vitellogenin genes progressively lost their function. The egg-laying platypus, which produces some yolk but not as much as birds and reptiles, only has one functioning vitellogenin gene. Marsupials and placental mammals, which don't rely on yolk at all to nourish their young, have none. The only evidence of vitellogenin in the marsupial and placental genomes is broken bits and pieces of these ancient and now defunct genes.

Vitellogenin loss could not have happened unless our premammalian ancestors figured out another way to feed their developing babies. Enter the casein genes. Some of the most abundant proteins in breast milk are caseins (also known as curds, as in the nursery rhyme "Little Miss Muffet"). Like vitellogenin, caseins are an important source of protein, but they also carry phosphorus and calcium, which are essential for embryo development. While casein proteins are only found in mammals, they belong to a much larger and more ancient family of proteins involved in calcium regulation in bones. So caseins probably first arose by the duplication and specialization of one of these ancient calcium-regulating genes. There are three casein genes in mammals, and *all* mammals have all three copies. This means that these important genes were present in the very first mammals that lived about 200 million years ago, but we think they were also present in some of our premammalian ancestors before mammals arrived on the scene.

What do we know about these premammalian ancestors? They laid parchment-covered eggs and kept their skin and eggs lubricated and waterproofed with oil and sweat. We know from the fossil record that the bodies and eggs of our premammalian ancestors became smaller as they became warm-blooded. We think the direct ancestors of mammals were nocturnal, probably ate insects, and were as small as mice or rats. Smaller animals lay smaller eggs, but a smaller egg might not have been able to pack enough yolk to support the development of these warm-blooded animals. The caseins might have helped solve this problem. If those sweat glands started secreting caseins, essential nutrients like calcium could have passed through the permeable eggshell to the developing embryo inside.

Once these ancient glands were producing a secretion with some nutrients for developing embryos in eggs, it's not a stretch to imagine that young hatchlings would have benefited from this secretion as well. Indeed, babies of our premammalian ancestors probably hatched earlier and were less developed as eggs became smaller. Initially, hatchlings would have simply licked secretions off the bellies of their moms, as the baby platypus still does today. Eventually, nipples evolved in the ancestor of marsupial and placental mammals. Nipples are a safer way to transfer milk to a baby. For the same reason that milk spoils if you forget to put it back in the fridge, a pool of milk on the mom's belly probably invited bacterial growth, which could have been dangerous for the baby drinking it.

Caseins are just one of the many unique ingredients found in breast milk, and Oftedal provides an explanation for how many of them may have first arisen. A protein that is essential in making lactose, one of the main sugars in milk, resembles an ancient antibacterial protein that was likely secreted from the skin glands of early amphibian-like land dwellers over 350 million years ago. The gene for that antibacterial protein was duplicated and then tweaked long ago (about 300 million years ago, like the casein genes). This may explain how milk sugars first arose. With nutrients like caseins and milk sugars passing to the baby directly from the mother's skin glands, there was less need for the vitellogenin yolk genes, and they were gradually lost.

This explanation for the evolution of the mammary gland is a simplification, and there are details I'm not including or that are not yet understood. For example, the apocrine sweat gland is a rather simple gland, so it would have needed to evolve a more complex tree-like branching pattern on its way to becoming a mammary gland. But the "numerous, successive" steps toward the evolution of lactation do not seem so inconceivable anymore.

To summarize, our breasts likely started off as simple sweat glands that lubricated and waterproofed the skin and eggs of our ancestors and kept them free of germs. After the duplication and specialization of specific genes, like the caseins, nutritive components were gradually added to the sweat, which were first passed to the baby in the egg and then directly to hatchlings. The evolution of nipples then made it safer for baby mammals to take their milk, which was delivered fresh and clean. Once the fundamental ingredients of milk evolved, individual mammalian species tweaked the recipe to meet the specific needs of their newborns.

Mammary glands and lactation were undoubtedly important players in the evolutionary success of mammals. But once this intimate exchange between mother and child evolved, the potential for genetic conflict emerged. I mentioned in the introduction that mothers and children don't have identical evolutionary interests because they don't share identical sets of genes. In chapter 7 on early motherhood, we will see that breastfeeding is one arena in which this conflict plays out.

Everlasting breasts

If I showed you a lineup of a hundred adult female mammals, you'd immediately be distracted by the more conspicuous differences among them. You'd notice that some mammals are tiny (the smallest is the bumblebee bat, which is one to two inches long and weighs a fraction of an ounce), and some are humongous (the largest is the female blue whale, at almost one hundred feet long and four hundred thousand pounds; on a related note, this whale mama produces 150 gallons of milk per day for her nursing calf!). You'd notice distinguishing traits like the elongated snout of the giant anteater, the orange and black stripes on the tiger, or the spines and bill of the echidna. The trait that we've been discussing at length in this chapter, mammary glands, would go unnoticed in my lineup. Except for a slight swelling when a female is pregnant and nursing, breasts are not visible on most mammals, let alone differences between the bosoms of different species. There is one exception—humans. The female breast is one of our distinguishing traits, like the anteater's snout or the tiger's stripes.

As discussed earlier, two features of the breast change during puberty—the milk-making equipment develops, and the fatty matrix develops. Development of the milk equipment is similar across species, especially within the placental mammals. It is the amount of fat laid down during puberty that is unique to humans. No other female mammal deposits so much fat that the nonlactating breast is visible (and in some cases apparently, difficult to avert your eyes from!). Large fat deposits are clearly not necessary for the lactating breasts to function properly; otherwise, we'd find them in all mammals. So why do human females put them there?

The variety of scientific explanations for the everlasting human breast may match the variety in human breast size itself. Our breast-obsessed culture seems to have infiltrated even the sciences (although, to be fair, not all cultures are equally fixated on breasts). Here is a sample of some of those hypotheses:

- Breasts evolved to look like buttocks, which encouraged human males to pursue sexual intercourse facing their partners rather than from behind, which contributed to bonding between males and females, which contributed to males playing a greater role in family life.
- The fatty breast evolved as an energy reserve to tap into later during nursing.
- As humans lost most of the hair on their bodies, breasts evolved to give babies something to cling on to during nursing (most primates cling on to their mom's chest hair while nursing).
- Fatty breasts evolved to advertise the genetic quality of potential mates to males.
- They evolved simply because males thought they looked good.

Most of these ideas don't have much data to back them up. There is no evidence that big butts evolved before big breasts (humans are also unique among mammals for having such rounded buttocks), so it is difficult to take that one seriously. The idea that permanent breasts evolved to store energy for a nursing baby makes intuitive sense, but no other mammal has arrived at the same solution, and they likely would have if fatty breasts were solving an important problem. Indeed, we know that fat from other parts of the body is the first to go when a female needs to tap into her fat reserves during breast-feeding. As for the breast as a baby handle idea, I don't know about the rest of you who have breastfed infants, but there is no way my nursing newborns could have possibly clung to my breasts (Ow!) without a football or cross-cradle hold or some kind of sling to keep them from falling off. This was likely true of human babies in the past.

The last two ideas on my list—breasts evolved to adver-tise female quality, and they evolved because they look good to men—are still being actively discussed in the literature. The idea that big breasts are a sign of female quality is one of many similar claims that specific traits (in humans and other animals) have evolved as honest signals of mate quality. In other words, certain physical and behavioral traits, like large breasts, elaborate plumage, and snazzy courtship dances, are adaptations that indicate to potential mates how intrin-sically valuable the individuals displaying those traits are. This is sometimes referred to as the "good genes" hypothesis on mate choice, which we will discuss in more depth later. Remember the bright red comb of the rooster we discussed

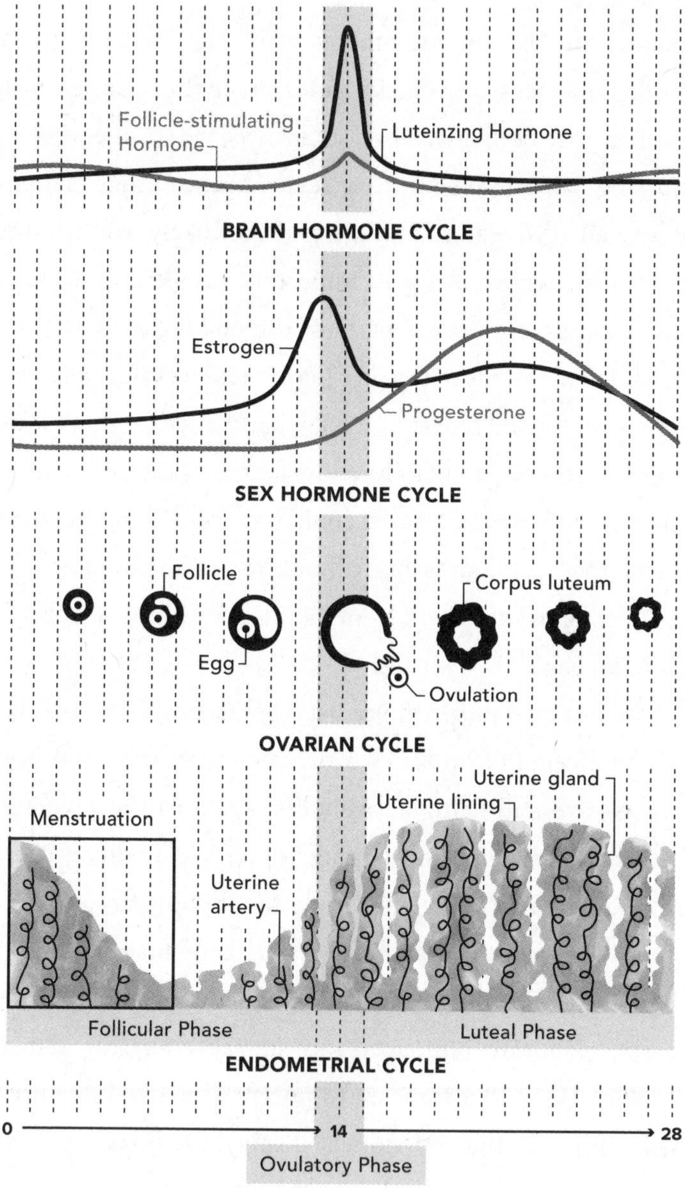

Follicle-stimulating Hormone

Luteinzing Hormone

BRAIN HORMONE CYCLE

Estrogen

Progesterone

SEX HORMONE CYCLE

Follicle

Egg

Corpus luteum

Ovulation

OVARIAN CYCLE

Menstruation

Uterine artery

Uterine lining

Uterine gland

Follicular Phase

Luteal Phase

ENDOMETRIAL CYCLE

0 ———————————————→ 14 ——————————————→ 28

Ovulatory Phase

Key events during the human menstrual cycle.

Illustration based on original by Merck & Co, Inc.

in chapter 1? Supporters of the good genes hypothesis would argue that the hen evolved a preference for the brightest, biggest comb because it signals to her that he is genetically superior to the one with the drab, small one. There is some evidence that roosters with brighter combs resist parasites better.

How would big breasts signal good genes in a woman? Proponents claim that breasts advertise to men which potential partners are best able to conceive children. The logic goes something like this: Women with higher hormone levels are more fertile; they also have larger breasts. So large breasts in females *and* the preference for them in males evolved because large breasts translate to more babies. One frequently cited study on 120 Polish women found a correlation between breast size and circulating estrogen levels (the higher the estrogen, the bigger the breasts). The authors of the Polish study didn't directly connect bigger breasts to higher fertility in the women they studied but instead used estrogen as a measure of their potential to have children. They cited a prior study that followed hormone levels of twenty-four women trying to get pregnant and found that levels of estrogen in individual women were higher in cycles in which they got pregnant than in cycles in which they did not.

Yale biologist Richard Prum offers a critical review of the Polish breast study and, more generally, a scathing rebuke of the good genes hypothesis and all those who continue to prop it up in his fascinating book *The Evolution of Beauty*. He complains that the authors of the Polish breast study provided no evidence that the slight differences in estrogen levels between large- and small-breasted women could have a significant effect on fertility, and he goes so far as to say that they falsify their own hypothesis by not directly correlating large breasts with more conceptions in the same women. While Prum may be overstating his case here—there are many reasons the authors may not have made the correlation—the relationship between hormone levels and fertility is far from clear in the literature. As I mentioned, there is some evidence that in individual women, cycles in which they conceive have higher estrogen than cycles in which they don't conceive. But it is well documented that absolute levels of reproductive hormones (estrogen and progesterone) are much higher in Western populations compared to others (probably because of fattier, more caloric diets as I'll discuss in the next chapter), and yet that doesn't translate to a higher potential of getting or staying pregnant in Western populations. Moreover, the relationship between estrogen levels and breast size is far from clear. In contrast to the study on Polish women, many other studies have failed to find any correlation between breast size and estrogen levels.

Prum offers a simpler explanation for the permanent female breast, and it draws on his experience as an ornithologist who studies the evolution of ornamental and behavioral traits in birds (I had the opportunity early in my graduate career to collaborate with Prum on a study of the evolution of feather and scale keratins in birds and reptiles). As with the exquisite plumage and elaborate courtship dances of many birds, Prum thinks permanent breasts evolved simply because the beholders of this trait—in this case, human males— thought they were beautiful. As Prum lays out in his book, the idea of aesthetic evolution by sexual selection was described by Darwin in his book *The Descent of Man, and Selection in Relation to Sex* to explain the exaggerated sexual ornaments of male birds. Darwin was puzzled, even made "sick"[1] by the intricate eyespots on the peacock. He grappled with what purpose the eyespot could possibly serve and how it could have possibly evolved incrementally over time. It was the peacock's eyespot and other elaborate, impractical ornamental traits that inspired his idea that animals have a "taste for the beautiful." According to Darwin, these *arbitrary* tastes drive the evolution of traits that have no function other than pleasing the beholder. And importantly, because of heredity, the beholder's children will also be considered beautiful by potential mates, which is how ornamental traits and the preferences for them become exaggerated over evolutionary time. As summarized by Prum, Darwin's ideas on aesthetic evolution by sexual selection were dismissed by his colleagues and then forgotten by his successors, who favored adaptive

---

1 Charles Darwin to Asa Gray, 3 April 1860, Darwin Correspondence Project, Letter 2743, https://www.darwinproject.ac.uk/letter/?docId=letters/DCP-LETT-2743.xml.

explanations for the evolution of animal ornamentation and mating preferences (like good genes). Prum's book is an effort to revive Darwin's original idea, providing his own documentation of these tastes for the beautiful in a variety of bird species and critically examining many studies, like the Polish breast study, that continue to be cited as evidence of good genes but don't hold up under scrutiny.

So according to Prum (and perhaps Darwin if we could ask him), large breasts aren't a sign of higher reproductive potential or genetic superiority—they exist because they were beautiful to males. We will explore the mechanism by which aesthetic evolution works in a future chapter. And on a related note, according to Prum, ancestral human males weren't the only ones with a taste for the beautiful. In his discussion of aesthetic evolution in humans, he provides an entertaining explanation for why the human penis is so large compared to the scrawny penises of our primate relatives—the larger penis both pleased the eye and pleasured the body of our female ancestors.

———

The verdict is still out on why the permanent breast evolved in humans. Although Prum makes an impassioned case for aesthetic evolution, none of the hypotheses to explain large breasts (and other attractive traits) have overwhelming support, and they are not necessarily mutually exclusive, as we will explore later. As a woman who is less endowed than many, I can't help but be offended by the notion that large breasts are a signal of genetic superiority (I'm very fertile by the way, with four strapping children!). I'll end our discussion of breasts by mentioning one of my favorites of Prum's arguments against good genes. For most of a woman's life, she regularly visits her ob-gyn to be prodded, tested, and questioned. Our gynecologists and obstetricians are experts in their field, with the most advanced medical knowledge and the most sophisticated medical equipment at their disposal. But until something goes wrong (like a woman can't get pregnant after years of trying), our doctors have no idea what our chances are to have children (and even if something does go wrong, they often have a difficult time figuring out why). Given how challenging it is for doctors to know and predict our reproductive potential and how challenging it is for scientists to find meaningful, repeatable connections between certain traits (like large breasts) and good genes, is it really plausible that prehistoric human men could have assessed the quality of a potential mate by simply looking at her cleavage? Perhaps they stared at breasts then for the same simple reason people do it now—because they look good!

CHAPTER 3

Is There a Point to Periods?

One of my favorite activities with my kids is cuddling up to read a book, whether it's the Seussian tongue-twister *Fox in Socks* with my preschoolers, the wacky and irreverent Wayside School stories with my grade-schooler, or more involved science and history books with my middle schooler. A book that works for all my kids, despite their nine-year span in ages, is called *A Child Through Time*. Each page is a beautifully illustrated rendering of what your life would have been like as a poor girl in ancient Rome, a samurai in training in twelfth-century Japan, or a cabin boy on a pirate ship in the 1700s. The book is targeted to tweens, but even my five-year-old son loves turning the pages. I wonder about his dark imagination as he fixates on his favorite illustration, which shows an Aztec girl in the fifteenth century being forced to inhale spicy smoke from a fire of burning chili peppers as punishment by her elders. My mind also wanders as I'm reading the book to my kids. How did the eleven-year-old boy from the New Kingdom of Egypt feel about working the fields by the Nile under the hot sun? (I know how my own Egyptian American sons would feel— miserable!) What did the European girl from the last ice age think was happening to her when she started bleeding every month? And more practically, how did she handle cramps and menstrual blood while going about her daily business of scraping mammoth skins, gathering wood for fires, digging up roots to eat, and keeping from freezing to death?

Menstruation has symbolic and ritual significance in many human cultures, so that girl from the last ice age—which ended about eleven thousand years ago—was probably given some explanation for why she was bleeding. She also may have used something to soak up the blood. Thousands of years later, we know ancient Egyptians made tampons from softened papyrus, and ancient Greeks used wood covered in lint. In contrast to these limited (and painful-sounding) options, girls in many parts of the world today have an overwhelming number of choices at the local drugstore. Girls are also now armed with biological information about menstruation that they pick up from school and other women in their lives. In puberty education, we learn that the lining of the uterus builds up every month in preparation for a baby. If you don't get pregnant, the lining and some blood are shed from the body.

But beyond this basic understanding of the mechanics of monthly periods, we aren't much more knowledgeable about *why* we menstruate than the ice age girl was. As with all traits we discuss in this book, answering the deeper why questions requires looking at the broader evolutionary context. Humans are members of a large group—placental mammals—who all have fairly similar reproductive biology. From elephants to skunks to rats to humans, female placental mammals make all the eggs they will ever have when they are a fetus, hit puberty at the end of childhood and start ovulating those eggs, have long pregnancies (relative to their cycle length), give birth to relatively well-developed offspring, and breastfeed their newborns. But of the five thousand or so species of placental mammals on the planet today, the vast majority do *not* menstruate. Only about eighty species, or 1 percent to 2 percent of all placental mammals, get periods. The rest are perfectly able to become pregnant and successfully bear children without menstruating. So why do human females have to experience the monthly onslaught of cramps and bleeding and in some cases more serious issues like heavy bleeding and endometriosis?

As a woman, I have naturally wondered about this, perhaps during a particularly difficult cycle or when I'm reading *A Child Through Time* with my kids. But I have also thought extensively about this question in a professional capacity. I earned my PhD studying the evolution of pregnancy and specifically the evolution of maternal tissues involved in pregnancy. One can't investigate these tissues without eventually asking the same question—why?

In this chapter on periods, I'll share some of the evolutionary explanations for menstruation, focusing on the only one that makes sense to me (full disclosure—it's one my colleagues and I proposed). First, I'll review *how* periods happen: What are the monthly hormonal and physical changes in our reproductive tissues, and how do they lead to menstruation during infertile cycles? Next, I'll tackle the *why*: What is the evolutionary significance, if any, of menstruation? Finally, I'll discuss how periods have evolved through human history, from the ice ages to today.

Up, down, and all around

In the last chapter, we discussed puberty, which kicks off when the brain starts sending hormones to stimulate the ovaries. Once the brain and ovaries start talking to each other, the outward changes of puberty ensue, including breast development and menstruation. Initially, as the reproductive system gears up, a girl's reproductive cycles are irregular and anovulatory (no egg is released), but after about a year or two, they recur more regularly. There is *much* variation in cycle length, both among women and in the same woman over the course of her reproductive life, with normal menstrual cycles ranging from twenty-one to thirty-four days. To keep things simple here, we'll look at a twenty-eight-day cycle, and we'll focus on what's happening in three parts of the body—the brain, the ovaries, and the uterus.

The most obvious benchmark of the reproductive cycle is menstruation, which starts on day one. The main hormone responsible for menstruation is progesterone, which is produced by the ovaries and maintains the endometrium—the lining of the uterus—in a receptive state. When progesterone levels drop in the last week of the cycle (we'll talk about why in a moment), the endometrium loses its hormonal support and the tissue starts to break down. This breakdown and shedding lasts roughly three to seven days.

While the uterus is shedding its lining in the first week of the cycle, what's happening in the brain and ovaries? They are gearing up for ovulation, which occurs about halfway through the cycle, on day fourteen. (Again, though, there is a huge amount of normal variation in the timing of ovulation, which is why the rhythm method of contraception—restricting intercourse to days of the menstrual cycle when ovulation is less likely to occur—is not the most reliable way to avoid pregnancy.) Ovulation is the release of an egg from one of the two ovaries, which remind me of small pomegranates engorged with seeds. In the ovary, the seeds are oocytes (or immature eggs), each encapsulated in a follicle, a sac that fills with fluid and contains additional cells that babysit the egg until it's ready to be ovulated.

The brain hormone that orchestrates the cycle is called gonadotropin-releasing hormone (GnRH), which is rhythmically released from the hypothalamus every one to two hours. In the first week of the cycle, GnRH travels to the pituitary gland and tells it to release follicle-stimulating hormone (FSH) and luteinizing hormone (LH) into the bloodstream. Rising levels of FSH stimulate the babysitting cells in the ovarian follicles to produce estrogen, and the follicles grow larger and accumulate fluid.[2]

Many follicles begin to develop in response to FSH, but one grows faster than the others. (In the case of fraternal twins conceived naturally, two grow faster than the others, usually one from each ovary.) The dominant follicle begins pumping out so much estrogen that it starts to inhibit the release of FSH from the brain, starving the smaller follicles of the hormone they need to continue development, and they die off. But the dominant follicle continues growing and releasing estrogen. At some point around day twelve or thirteen of the cycle, estrogen levels reach a threshold, which signals to the pituitary to release a whole lot of LH. This surge of LH triggers the last steps of dominant follicle development, including the rupture of the egg from the follicle and ovary around day fourteen. Some women experience cramps in their abdomen during ovulation, called mittelschmerz, which is German for "middle pain."

Rising estrogen from the ovaries in the first half of the cycle doesn't just signal to the brain. Estrogen also tells the

---

2   Our bodies actually make three main types of estrogen: estradiol (E2), which is the predominant and most potent form of estrogen made by the ovaries during our reproductive years; estriol (E3), which is made by the placenta during pregnancy; and estrone (E1), a much weaker estrogen made in the ovaries, adrenal glands, and fat cells. Unless otherwise specified, I'm referring to the potent form of estrogen, estradiol.

uterus to start repairing itself. After about a week of bleeding, the uterus needs to build itself up again. Estrogen is the initial messenger, directing endometrial cells to divide and make more of themselves. But in the days around ovulation, estrogen levels plunge and progesterone takes over, produced by the babysitting cells left in the follicle after ovulation (this ovarian structure is called the corpus luteum). In contrast to estrogen, which sends proliferative messages to the uterus, progesterone sends transformative ones, telling the endometrium to transform into a structure that can support a pregnancy. This includes transforming the most abundant type of endometrial cell—fibroblasts—into decidual cells, which look and behave very differently from their precursors. There is also an influx of immune cells into the uterus and extensive branching of blood vessels.

As all this is happening, the ovulated egg is traveling down the reproductive tract. If the cycle is a fertile one, the egg will get fertilized on its way down, will divide several times to form a ball of cells, and will be ready to attach itself to the endometrium about ten days after ovulation. If fertilization happens and all goes well, a hormone made by the fertilized egg (human chorionic gonadotropin, or hCG—the one detected in an over-the-counter pregnancy test) will tell the corpus luteum to continue making progesterone. But if fertilization doesn't happen, the corpus luteum withers away in the fourth week of the cycle and stops making progesterone. Without progesterone, the transformed endometrium falls apart. Decidual cells starved of progesterone die and are shed from the uterus, just as deciduous trees shed their leaves in the winter. Because decidual cells keep the recently elaborated blood vessels healthy and intact, decidual cell death also results in some local bleeding. The dead endometrial tissue and blood begin their exit from the uterus and body, and now we're back to the beginning. The cycle starts all over again if a pregnancy hasn't occurred.

The evolutionary purpose of periods

That background on the main events of the human menstrual cycle positions us to discuss why we need to have periods at all.

The history of scientific thought on menstruation goes back to Aristotle, who suggested that menstrual cycles were controlled by the moon and that menstruation was a means to rid the inactive female body of unused nourishment. A few others chimed in on the topic over the centuries, with similarly uninformed explanations (the classical Greeks, for example, didn't dissect human bodies so had no knowledge of where menstrual blood even came from).

The modern conversation about why menstruation evolved started in the 1990s. Most anthropologists and biologists have tried to ascribe an adaptive function to menstruation, suggesting that cyclical bleeding must be evolutionarily advantageous to females, increasing the odds of reproductive success. Biologist Margie Profet put out the first idea in the early '90s, suggesting that menstruation evolved to rid the

uterus of viruses, bacteria, and parasites that make their way
into the female reproductive tract during sexual intercourse.
The community heavily criticized Profet, as scientists often
do in response to new and provocative ideas. The controversy
continued until anthropologist Beverly Strassmann explicitly
tested Profet's hypothesis. If Profet was right and menstrual
blood expels pathogens from the body, you'd expect fewer of
them after menstruation than before. Among different species,
you'd also expect more uterine pathogens and more menstrual
blood in species that are more promiscuous. Strassmann
looked at the existing data, which did not support any of these
predictions. In fact, there appear to be fewer microorgan-
isms in our reproductive tract right *before* menstruation, the
opposite of what you'd predict.

Strassmann didn't just test Profet's idea; she also offered
one of her own. She argued that menstruation evolved to save
energy. Keeping the endometrium in the transformed state
of pregnancy requires more energy, Strassmann said, than
shedding it and rebuilding it every cycle. Strassmann made
some calculations about how much energy is saved—she
estimated that over four months of cycling, a woman saves the
amount of energy in six days' worth of food.

That all sounded convincing until biologist Colin Finn
pointed out a basic problem with Strassmann's hypothesis. If the endometrium remained in the transformed state
of pregnancy indefinitely, a woman wouldn't be able to
get pregnant at all. The reproductive tract must perform

many different functions during the reproductive cycle and pregnancy, including transporting sperm up and ovulated eggs down. The uterine conditions needed for sex cell transport and activation (lots of uterine fluid) are very different from those needed during embryo implantation and early pregnancy (little fluid). So maintaining a transformed endometrium with little fluid in the uterus would make it difficult for sperm and egg to ever find each other, which is likely why we don't observe this trait in nature. Strassmann's hypothesis didn't adequately address what's going on in nonmenstruating species, which comprise the vast majority of placental mammals—they don't menstruate, but they also don't sustain a transformed endometrium. The general problem with many of the explanations for menstruation is that they don't tell us why menstruating species have periods but all other mammals do not.

So when my colleagues and I started thinking about the significance of menstruation, the first thing we did was take a thorough look at which other placental mammals have periods. Chimpanzees, gorillas, and some monkeys menstruate, which is not surprising since they are our closest animal relatives, and we often share traits with those to whom we are closely related. What did surprise us, though, is the random set of nonprimate mammals that do menstruate: a handful of bats, the elephant shrew (which is related to elephants), and a very recent addition to the collection—the Egyptian spiny mouse, the only rodent known to menstruate. An important point is that these species are not closely related to each other, and being located in very different parts of the mammalian family tree, they are more closely related to species that do not menstruate than they are to each other. And by the way, many people have asked me about bleeding in dogs, which is often mistaken as menstruation. In dogs, bleeding occurs *before* ovulation, not after, and it happens because some blood cells leak through the walls of small blood vessels in the uterus when estrogen levels are high. This is a very different phenomenon than menstruation, which is the shedding of blood and endometrial tissue triggered by a drop in progesterone.

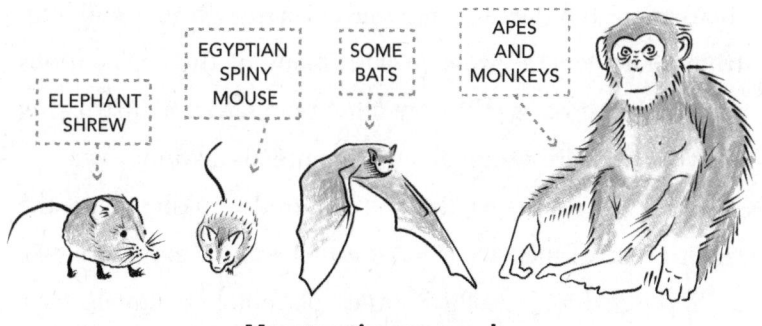

Menstruating mammals.

This unusual distribution of menstruating mammals is useful for several reasons. First, when a trait pops up multiple times in different animals, it usually indicates that the trait is adaptive. Flight evolved independently in birds, bats, insects, and extinct pterosaurs, which supports the obvious conclusion that flight is an adaptation in these groups. We think menstruation evolved at least four times—once in primates, once in rodents, once in the elephant shrew, and at least once in bats—which suggests that it has a specific purpose. Otherwise, why would it have emerged so many times? We get another clue by looking at other traits these animals share. Is there anything else we all have in common?

The answer is most definitely yes. We were interested in one trait in particular. You've likely never heard of the jargony name for it—spontaneous decidualization—but if you read the last section, you have probably already figured out what it means. We discussed what happens to the endometrium during the human menstrual cycle—first estrogen builds it up, and then progesterone transforms it. Endometrial cells transform into decidual cells to get ready for pregnancy. This transformation is spontaneous decidualization.

Spontaneous decidualization does not occur in most mammals. In most mammals, the transformation does happen, but only *after* a female gets pregnant. We refer to this as plain ol' decidualization, or sometimes "induced" decidualization. In nonmenstruating mammals, the *baby* induces the transformation. This makes intuitive sense—invest the energy to transform your uterus only when you are actually pregnant.

But in menstruating mammals, the uterus transforms "spontaneously," regardless of whether a female is pregnant or not, driven by progesterone and other signals from mom. In menstruating mammals, then, the *mother* induces the transformation. It happens very predictably at the same point in every cycle, and it does not require an implanting embryo.

So all menstruating mammals—and *only* menstruating mammals—transform their uterus every cycle in preparation for a baby. Biologist Colin Finn first noticed this correlation, and he also suggested a causal connection between the two traits. Once the endometrium transforms, if its lifeline—progesterone—is removed, tissue breakdown and bleeding must occur, whether it's at the end of an infertile cycle in menstruating mammals or at the end of pregnancy in all placental mammals. As we discussed above, the decreasing level of progesterone in infertile cycles is the immediate trigger for menstruation in women. Similarly, a termination of progesterone signaling at the end of pregnancy in all placental mammals results in birth, which includes a shedding of endometrial tissue and blood (so much blood!—something first-time mothers are not prepared for). A colleague of mine, Mihaela Pavlicev, has argued that the uterine activities during human birth (inflammation, contractions, and shedding) are nothing more than a nine-month delay of menstruation. Many experiments on mice and on cells in the lab support the idea that in mammals, endometrial breakdown and bleeding are inevitable consequences of decidual transformation followed by a drop in progesterone.

With this information, then, it is clear that we've all been asking the wrong question. The trait in question is not menstruation, which is just a side effect of the spontaneous transformation of the endometrium followed by a drop in progesterone. What we should be asking is, why do we transform our endometria every month when most mammals don't?

The honest answer is that we don't know yet. But we do have two promising leads, both of which point to the evolutionary negotiations between mother and child that I discussed in the introduction. Let's review these negotiations in brief before diving into the leads.

The evolutionary interests of close family members are often the same. It is in a mother's interest to feed and protect her children, because these behaviors increase the likelihood that her children will survive and pass on her genes. It is in the child's interest to seek nourishment and protection from its mother, because these behaviors increase its own chances of survival and reproductive success. However, the evolutionary interests of mother and child are *not* identical. A mother is equally related to all her children, whereas a child is only 50 percent related to its mother and full siblings, 25 percent related to half siblings, and 100 percent related to itself. According to the conflict theory, this setup results in greedy children and stingy mothers. Genes in children that help them take more resources from their mothers will fare better over evolutionary timescales than those that don't, and genes in mothers that help them moderate access to those resources will fare better than those that don't. In today's world, where nutrients are abundant and easy to come by for many of us, this scenario doesn't make much sense, but in the evolutionary past, nutrients were scarce and would have been a source of conflict for competing individuals.

There is evidence of this conflict in many places and at

many levels, and the uterus is one of its major arenas, because it is the site where resources are directly transferred from mother to child during pregnancy. We think of the uterus as a cozy place for a baby to develop, but you can also think of it as a battleground. What are the "weapons" used by mother and child at this uterine battleground? On one side, the child produces placental cells and molecules that better invade uterine tissues and extract nutrients from the mother. The placenta is made by the child—it has the same genome as the developing baby—and the cells of the placenta take on the role of interacting with the maternal tissues of the uterus and establishing the conduit between mother and child during pregnancy. From the child's perspective, it wants the conduit to be as big and generous as possible.

On the other side, the mother makes uterine cells and molecules that resist excessive invasion. The maternal decidual cells we've been talking about are multitaskers, producing thousands of molecules that are involved in the immune response and in embryo implantation but also, critically, in placental restraint. In contrast to their untransformed precursors, decidual cells of the uterus physically resemble a wall of shields on a frontline. Another evocative metaphor I've come across is that the uterine lining changes from soil to bricks during its transformation. Decidual cells permit the invasion, but they also keep placental cells from going too far. One way we know this to be true is by observing how pregnancies unfold at sites without a proper (or any) decidual transformation, as in the fallopian tubes or on a C-section scar. Because these sites lack decidual cells, placental cells invade without restraint, which can result in maternal hemorrhage and even death.

At the geopolitical scale, the way a conflict between nations unfolds depends on the decisions made by each party at different stages of the conflict. Some wars escalate, some result in quick surrender, and some reach a diplomatic solution that appeases all parties. Historical contingency is also true of negotiations at the genetic level, for which we see evidence when we observe how they unfolded and were resolved in the uterus/placenta of different mammals. In some mammals, like horses and cows, the mother's genes appear to have gained the upper hand, and the placenta is strictly prevented from invading endometrial tissue during pregnancy (so nutrients and oxygen have to diffuse across many tissue layers to get to the baby). This setup is safer for a mother, and her endometrium never needs to transform into a protective shield, since the fetus's tissues never breach maternal ones. At the other extreme, the ancestral ape placenta became hyperinvasive, evolving a new type of placental cell that invades maternal blood vessels in a novel way. Placental cells of the human fetus burrow all the way through the endometrium into muscle layers of the uterus, aggressively remodeling maternal blood vessels for greater access to maternal resources.

All this brings me to the first hypothesis on the evolution of spontaneous decidualization (and as a by-product, menstruation). In the primates, as a possible countermove to an increasingly invasive ape placenta, spontaneous decidualization may have evolved to protect the mother in advance of the invasion. As one of the functions of the transformed uterus is to restrain invading placental cells, this countermove may have offered some early uterine protection to the mother. Instead of waiting for signals from the baby, the mother took back some control over her uterus, starting the transformation earlier and all on her own. Back to the idea of historical contingency, this early transformation may have been advantageous in menstruating species but not in nonmenstruating species, because only menstruating species have such an invasive placenta. (Also, the set of mutations enabling the placenta to become more aggressive in the first place only occurred in some species and not in others.) In the primates, there is a correlation between the degree of placental invasiveness, the extent of endometrial transformation during the menstrual cycle, and the amount of menstrual blood that is shed. Humans represent the most extreme case on all fronts. But these comparisons are mostly anecdotal, with no formal comparisons made in primates or in the other groups with menstruating species. These tests are difficult to perform, as many of the relevant animals are not lab models.

Paradoxically, recent lab experiments showed that human endometrial cells grown in a flask are more *permissive* to placental invasion compared to those of other mammals, a fascinating observation I'll return to in chapter 8 on cancer. As placental cells became more invasive in the ancestor of apes,

it appears that uterine cells evolved to be more accommo-
dating of the invasion. Some might argue these results falsify
the hypothesis presented above. While this is possible, I think
the results are a reminder of the complex role of the endometrium in both fetal accommodation *and* restraint. David Haig,
the evolutionary theorist at Harvard who first described the
conflict of interest between mother and fetus, has written that
conflict and cooperation are two sides of the same coin. Once
the ancestral ape fetus "decided" to become more invasive,
the best evolutionary response in the mother may have been
to transform the uterus early to have more control over the
invasion while also allowing the fetus greater access to oxygen
and nutrients.

The second idea on why spontaneous decidualization
and menstruation evolved, which is not mutually exclusive
with the first idea, also suggests it was a maternal move in the
evolutionary negotiations between mother and child. In this
model, the early transformation evolved as an embryo screening strategy, enabling the mother to quickly weed out poor-
quality embryos. There is now ample evidence that the female
reproductive tract acts as a quality control device, screening
for the best eggs, sperm, and embryos. Human embryos have
a high rate of chromosomal and developmental abnormalities, with an estimated 40 percent to 60 percent of all fertilizations ending in a loss. Our unusual sex lives are to blame
for some of this. While most mammals only copulate in the
short window around ovulation, humans engage in sexual

intercourse throughout the reproductive cycle. As a consequence, many fertilizations in humans happen with eggs or sperm that are too old, resulting in embryos that won't develop normally.

Do these impaired embryos simply die early and pass through on their own, or does the reproductive tract actively discard them? There is experimental evidence from cultured cells in the lab that shows that decidual cells, but *not* their untransformed precursors, trigger a strong chemical response against low-quality embryos. In a newly pregnant woman, that would result in an early miscarriage, for decidual cells would prevent the embryo from implanting into the uterus. An early miscarriage can be distressing (I had one myself), but from the evolutionary perspective, it is preferable to the alternative—a later miscarriage or the death of an infant or child. Clinical data show that women whose endometria don't transform properly in the second part of the menstrual cycle become pregnant at a high rate, but they also have a high rate of miscarriage later in their pregnancies. Because their endometria aren't effectively screening embryos, these women are investing months in embryos that can't develop to term.

The early embryo screening process in menstruating species can be viewed as an adaptation for mothers to deal with so many low-quality embryos. Taking control over the process of endometrial transformation (rather than leaving it to the implanting baby) may have boosted a mother's reproductive success because earlier embryo screening saved time and energy. But as we've discussed, children are not passive objects but active agents with their own agendas. According to the conflict theory, genes in mothers and genes in children have different criteria for continuing a pregnancy. Mothers want a sure thing, preferring to invest in embryos from which they are most likely to reap their evolutionary rewards. Embryos, on the other hand, will adapt to stick around as long as possible, evolving molecular smoke screens to trick the mother into letting the pregnancy stick. We don't know what these smoke screens are (nor do we know precisely how the endometrium does its weeding out), but the fact that some babies are born with genetic diseases suggests that smoke screens exist or that maternal screening isn't perfect.

Consistent with the embryo screening model, many menstruating mammals mate throughout the reproductive cycle, have small litters, and invest heavily in each developing child. When the cost of making/raising a child is especially high and when the rate of abnormalities is high, we would predict the evolution of more effective and earlier screening mechanisms. In addition to humans, rhesus monkeys—another menstruating species—have been shown to have a high rate of chromosomal and developmental abnormalities in early embryos. If this was found to be true in other menstruating species, it would support the idea that spontaneous decidualization evolved to screen out low-quality embryos, which may be more prevalent in species that menstruate because of extended mating.

Whether spontaneous decidualization (and menstruation as a by-product) evolved to weed out unhealthy embryos or to mount an early defense against invasive offspring, both suggest a sobering view of the mother-child relationship. The evolutionary compromise that we observe today in humans isn't an ideal setup for either mother or child, but because of the particular way the conflict escalated in our ancestors, this is what we're stuck with. Beyond the inconvenience of menstruation, we are vulnerable to many disorders of the reproductive tract as a consequence of the conflict. Endometriosis, which affects 5 percent to 10 percent of women, is a painful disease characterized by the growth of endometrial tissue outside the uterus. The disease is only found in menstruating species because the movement of tissue during menstruation can lead to endometrial cells reaching sites like the fallopian tubes, ovaries, and pelvic cavity. As we will discuss in chapter 6 on pregnancy, mother-child conflict is also invoked to explain why we suffer from preeclampsia, gestational diabetes, and other pregnancy-related complications.

It might be tempting for readers to think that the human fetus is more advanced than other mammals in evolving a more invasive placenta. Some have argued that the aggressive human placenta enabled the evolution of large brains in our species, but we know that a mammalian fetus can grow to be quite large without being invasive at all. Whales have the least invasive type of placenta, but they have the largest brains on earth. As in battles at the geopolitical scale, then, it is not always clear how much the "winner" in mother-child conflict has gained, and a different strategy may have led to a similar outcome without the costs. Those of us who endure monthly periods, who carry the battle scars of this age-old conflict, certainly understand those costs.

Periods through time

Let's return to our adolescent girl from the last ice age. When we talk about menstruation, we can't help but think of our own experiences and those of women in Western countries today who practice birth control. The average American woman has on the order of four hundred reproductive cycles in her lifetime, the vast majority of which do not end in a pregnancy. But what about women without access to birth control, like the ice age female? How many periods did she get in her lifetime, and how did she experience those periods?

We know that women in the past (and still today in some populations) spent most of their reproductive life either pregnant or breastfeeding. Since you do not ovulate when you are pregnant or nursing frequently, the number of reproductive cycles in women who don't practice birth control is about one hundred, as Beverly Strassmann quantified in her work with women of the Dogon tribe of Mali. Many biologists and doctors have commented on the health consequences of this difference. Whether or not you have children, it is important to know that our bodies evolved to cycle about one hundred times, not four hundred. One consequence of having more infertile cycles in our lifetimes is a higher risk of breast cancer. Remember that during every cycle, our reproductive tissues prepare for a pregnancy by proliferating and then transforming. We discussed this in the context of the uterus, but proliferation also happens in the breast during each cycle in response to estrogen. Every time a cell divides, there is an opportunity for a cancerous mutation to arise. The result of more cell divisions is a breast cancer rate in women from North America and Western Europe that is many times higher than that in women who don't practice birth control. You would expect an increase in breast cancer risk with birth control methods that don't involve hormones (e.g., condoms), but hormonal birth control (e.g., the pill) complicates the story because hormones act on many tissues in the body, and there are a variety of hormone formulations on the market that may influence cancer risk differently, something being actively investigated by Strassmann. Overall, studies show that oral contraception increases the risk of breast and cervical cancer but decreases the risk of endometrial, ovarian, and colorectal cancer. A high percentage of women use hormonal birth control, so clearly more research is needed.

Anthropologist Wenda Trevathan has written extensively about this issue in her book *Ancient Bodies, Modern Lives*. Her thesis is that many of the health challenges women face today, like a high risk of breast cancer, result from a mismatch between our evolved bodies and modern lifestyles, a topic we'll return to in chapter 8. You've probably heard about this mismatch in the context of diet and obesity. Modern diets—which for many are high in sugar, fat, and total calories—look nothing like the diets of our human ancestors. Trendy dieting strategies like the paleo diet or intermittent fasting attempt to replicate how ancestral humans ate, which is purported to be a better match for our evolved bodies. Modern diets also seem to affect our hormones and reproductive cycles, not just our waistlines. Trevathan brings up a couple of points in this regard.

First, richer diets and lower levels of physical activity are thought to contribute to higher circulating levels of ovarian hormones. In addition to a greater number of lifetime cycles in women who practice birth control as I just described, the average American woman has higher circulating estrogen and progesterone during each cycle than the average woman, for example, from the !Kung San tribe of Botswana. This does not appear to affect fertility, but higher hormone levels do affect other aspects of our reproductive health and well-being. They may fuel reproductive cancers as touched on above, and they may contribute to premenstrual syndrome (PMS) and longer periods of bleeding during our cycles. PMS, a catchall for any symptom experienced in the few days before menstruation starts—including breast pain, mood swings, and anxiety—is more common among women living in Western countries. As Trevathan and others have hypothesized, PMS might be a consequence of higher ovarian hormone levels in these women, which in turn might result from richer diets and sedentary lifestyles. The higher the rise in progesterone levels in the second half of the menstrual cycle, the greater the fall at the end of the cycle, and this hormone "withdrawal" may lead to the PMS symptoms that many of us experience. Higher progesterone levels might also be responsible for another phenomenon observed in women living in the United States and Europe—longer periods. Studies of women living in developing countries have shown that women menstruate only three or four days per cycle, compared to those in the United States and Western Europe, who bleed closer to six or seven days.

In my own work on menstruation, my colleagues and I presented a hypothesis on *how* spontaneous decidualization and menstruation evolved. It hinges on how the endometrium responds to progesterone in menstruating versus nonmenstruating species. A natural extension of our hypothesis is what should happen in women who have higher circulating levels of progesterone. You would expect a more extensive transformation of the endometrium during the menstrual cycle and, as a consequence, a longer period of shedding the additional decidual tissue and blood. Longer periods are exactly what we observe in women in the United States and Europe.

The second important issue regarding diets and periods is that modern diets may also contribute to a noticeable trend over the last one hundred to two hundred years—earlier menarche. Menarche, or the first menstrual period, happens in girls today around the age of eleven or twelve, but if you could ask your great- or great-great-grandmother when she had her first period, she would likely answer with a number many months or even years older than twelve. Early puberty is a hot topic right now among doctors and public health experts. While it is clear that girls reach puberty earlier now than in the recent past, there is much debate about why and whether it is cause for concern.

One factor involved in this trend is diet. Childhood nutrition and overall health have improved in the last two centuries, and this has likely advanced the age of puberty in girls. Many studies show that healthy, well-nourished girls get their first periods earlier than unhealthy, stressed, or undernourished girls. We know how some of this works. Nutritional status has a direct effect on the brain, stimulating GnRH, FSH, and LH secretion. Since the onset of puberty involves the activation of the ovaries via these brain hormones, diet can influence the timing of puberty. Consistent with this is the early onset of puberty in overweight and obese girls.

According to evolutionary theory, earlier menarche in better-fed girls is exactly what is supposed to happen. Our bodies are built to be somewhat flexible—if conditions are favorable, do X; if they are unfavorable, do Y. For most traits, our genes don't determine the precise expression of those traits, but rather they set the bounds of what is possible depending on environmental conditions. Human height is a simple example. We know height is influenced by our genes—very short parents are unlikely to produce a child destined for the NBA. That said, depending on your nutritional status as a kid, you may or may not reach the upper bounds of what your genes prescribe. Plenty of data show that kids born to migrant parents who moved from a developing to a developed country are taller than their parents. The same is true of puberty. While the timing of puberty has a genetic component, the environmental conditions you are exposed to both in utero and during childhood influence whether you enter puberty earlier or later. Those same taller girls with shorter parents from developing countries also hit puberty earlier than their moms, at least in part because of better nutrition in their new country.

But many are concerned that the trends we are observing now are not within the normal range. Our bodies evolved to operate within certain energetic limits that don't appear to exist now in some populations. Richer, bigger diets play a role in earlier menarche, and earlier menarche has been associated with a higher risk of diabetes, metabolic disease, heart disease, and certain cancers. While the reasons behind these associations are not well understood, one of the links could be childhood obesity, which influences both puberty timing and the risk of many diseases later in life. Moreover, just as girls' bodies are maturing earlier, it is taking them longer to gain the skills and knowledge necessary to function as adults in an increasingly complex and stressful world. This mismatch between our reproductive and psychosocial maturation may be occurring for the first time in human history, and it places a novel sort of stress on those who are physically mature but not yet socially or professionally mature or financially independent.

Another possible cause of earlier puberty is increased exposure to environmental toxins. Some argue that a change in nutrition can only partially explain the trend of earlier puberty and that something else must be going on. Endocrine disrupting chemicals (EDCs) are toxins that mimic and/or block the action of our reproductive hormones and are found everywhere—in our food, cosmetics, and personal care items. Polychlorinated biphenyls (PCBs), bisphenol A (BPA), and phthalates are well-known EDCs. They have been implicated both in reproductive disorders like infertility and miscarriages and in breast cancer. Several studies have looked at the effects of BPA on the timing of puberty with equivocal results. A few have found a connection between BPA levels and very early breast development (in girls under four years old) or precocious puberty (which is a clinical diagnosis of puberty in girls under eight), but no study has found a correlation between BPA levels and puberty timing in "healthy" girls. To complicate matters, BPA is thought to influence metabolism, and BPA levels are higher in obese girls. BPA has a short half-life in the body, and one possible explanation for differing results across studies is the different methods used to measure BPA and difficulties in obtaining accurate measurements. Clearly, more research is needed, for these toxins are ubiquitous in our environments, and early puberty is associated with adverse health outcomes in adult life.

Back to our ice age female. In some respects, she had it pretty good—she certainly wasn't exposed to EDCs, and her diet was likely a better match for her body than modern diets are now for many women. Diet and physical activity may have had far-reaching consequences, including when she got her first period, if and how she experienced PMS, how painful and inconvenient her periods were, and as we'll discuss in a later chapter, how she experienced menopause. She may not have understood why she periodically bled (which hopefully you do after reading this chapter), and she may not have had the best hygiene products for her periods. But she had many fewer periods than we do today (and therefore was at much lower risk of breast cancer, which would have been relevant only if she survived long enough to get it). That said, she was pregnant or nursing most of her adult life, which is certainly not going to become the new fad lifestyle! But trying to minimize exposure to EDCs and maintaining healthy diets and lifestyles are reasonable ways of tipping the scales toward better health outcomes for ourselves and for our daughters.

CHAPTER 4

On the Origin of Orgasms

In the iconic orgasm scene from *When Harry Met Sally...*, Sally, played by Meg Ryan, tries to convince Harry (Billy Crystal) that women fake it all the time. The scene unfolds in a crowded New York City deli over piled-high pastrami sandwiches:

> **Sally:** Most women at one time or another have faked it.
>
> **Harry:** Well, they haven't faked it with me.
>
> **Sally:** How do you know?
>
> **Harry:** Because I know.
>
> **Sally:** Oh, right. That's right. I forgot. You're a man.
>
> **Harry:** What is that supposed to mean?
>
> **Sally:** Nothing. It's just that all men are sure it never happened to them. And most

women at one time or another have done it,
so you do the math.

Harry: You don't think that I could tell
the difference?

Sally: No.

Harry: Get outta here.

Sally: Ooo...Oh...Ooo.

Harry: Are you okay?

Sally continues with a resounding performance, Harry holding on to the delusion that he's never been faked on until the climax of her show. While the other patrons in the deli are rapt and even a little titillated, Harry is left stupefied.

The female orgasm doesn't just confound men like Harry in this classic movie scene. A biological explanation for the female orgasm has haunted philosophers, anthropologists, and biologists for centuries. What is so puzzling about it from a biological standpoint? On the one hand, in contrast to male orgasm, female orgasm during vaginal intercourse doesn't happen easily (or at all) for many women, which is perhaps one of the points of Sally's performance. It is also not required for reproduction—many a fertile woman has never had an orgasm during vaginal intercourse. On the other hand, the intense and repeated pleasure the orgasm can give a woman involves a sophisticated physiological response that seems too complicated to have evolved for no reason at all.

The scientific discussion over the female orgasm is

one of the most heated and contentious in biology, and until recently, you could group the debaters into two main camps. In one camp are those who argue that the orgasm is in fact a reproductive adaptation for women. In other words, despite its fickle nature, orgasm still serves some important function that enhances a woman's reproductive success. A number of biologists have put forth some creative ideas on what that function might be. The other camp maintains that the female orgasm is a kind of biological accident. During early embryonic development, the clitoris in females and the penis in males start to form in the same place from the same tissues. There is no disagreement over why males have orgasms—since they are tied to sperm release, they evolved to deliver sperm to females. The logic, then, is that the clitoris and penis are derived from the same tissue, so women have orgasms because men require them later in life. They are simply a by-product or lucky accident, with no reproduction-enhancing function in women.

In this chapter, we'll take a closer look at these two sets of explanations for female orgasm, but we'll also focus on a very recent account that doesn't quite fit into either camp. Günter Wagner, who was my PhD advisor at Yale, and Mihaela Pavlicev, who is a colleague, assert that previous ideas have been so narrow in evolutionary scope that the correct evolutionary explanation of the female orgasm has been overlooked. To understand Wagner and Pavlicev's hypothesis, which now has some experimental support, we'll examine the broader context of female mammalian reproductive biology while keeping in mind the many nuances of the evolutionary process that we discussed in the introduction. According to Wagner and Pavlicev, the mystery of the female orgasm can be solved by decoding the evolution of another big O in our mammalian ancestors—ovulation. I think they're on to something, although their explanation might only be a starting point. But first, the basic biology.

What is an orgasm?

Many readers might be experts on how to achieve an orgasm and perhaps, like Sally, how to fake one too. But for a scientific understanding of this complex phenomenon, we'll need to examine the place in the body in which orgasms initiate, the clitoris, and then discuss the physiology of the orgasm itself.

Recall from chapter 1 that the future genital region of both sexes looks identical until about eight weeks of embryonic development. At this point, the ovaries and testes have just started producing hormones. It is clear that testosterone made by the testes directs development of a penis instead of a clitoris. In females who aren't making large amounts of testosterone (and who are making estrogen and responding to estrogen from their mothers), a clitoris develops. While the clitoris and penis sprout from the same region in the embryo and are molded from the same tissues, there are obvious differences between the fully formed structures. The penis becomes a large external organ within which the urethra—the tube that carries urine from the bladder—is enclosed. In contrast, the clitoris in humans is much smaller (at least the externally visible portion, see page 90), and it does not enclose the urethra. Also, while the head of the penis becomes enclosed in foreskin, the head of the clitoris— called the glans clitoris—is only partially covered by the clitoral foreskin, or hood. However, there is much variation across women, ranging from hardly any coverage of the glans to fuller coverage.

A noticeable difference between the clitoris and penis is that most of the clitoris is not externally visible. The anatomy of the clitoris had not been widely known until recently, perhaps because a male-dominated academia was not interested or too prudish, especially in the United States. While some European texts from the nineteenth century did accurately describe clitoral anatomy—one text published in 1844 by German anatomist Georg Kobelt is on *Wollust Organe*, which roughly translates to "lust organs"—American textbooks in the nineteenth and twentieth centuries either ignored the clitoris completely (the first editions of the well-known *Grey's Anatomy* did not mention the clitoris at all) or inaccurately described it as a small cylindrical structure composed mainly of the glans clitoris.

It wasn't until the late 1990s and 2000s, with the publication of work by Australian urologist Helen O'Connell, that clitoral anatomy became more widely recognized. O'Connell performed microdissections of cadavers and did MRIs on living women and published a paper in 2005 revealing a detailed anatomy of the clitoris. It reminds me of a tropical flower, so striking that it has inspired many artistic works, including a sculpture by Australian artist Alli Wolf called the *Glitoris*, a 100:1 scale model of a clitoris bedecked in gold and sequins. As the *Glitoris* shows, the structure has a few distinct parts: the glans clitoris, which is that pea-sized bit of tissue located between the front part of the labia and partially covered in its hood; a short body; two long crura (which means legs in Latin; together they resemble a wishbone); and two large bulbs. O'Connell has described the glans clitoris as the tip of an iceberg, with over 90 percent of the clitoris (the body, crura, and bulbs) tucked away in the pelvis. The crura and bulbs wrap around the front part of the urethra and vaginal wall and extend out to the thighs. While these were previously described to have a length measured in milli-meters, we now know that these dimensions are an order of magnitude off, with the adult clitoris measuring about five *inches* long. Moreover, the clitoris expands dramati-cally during sexual arousal, as we'll discuss below. Many have argued that the illusive and controversial G-spot in the vagina is just the internal parts of the clitoris that are stimu-lated through the vaginal wall during intercourse. In fact, O'Connell and others have shown that there is no anatomical evidence of a G-spot in the vaginal wall itself.

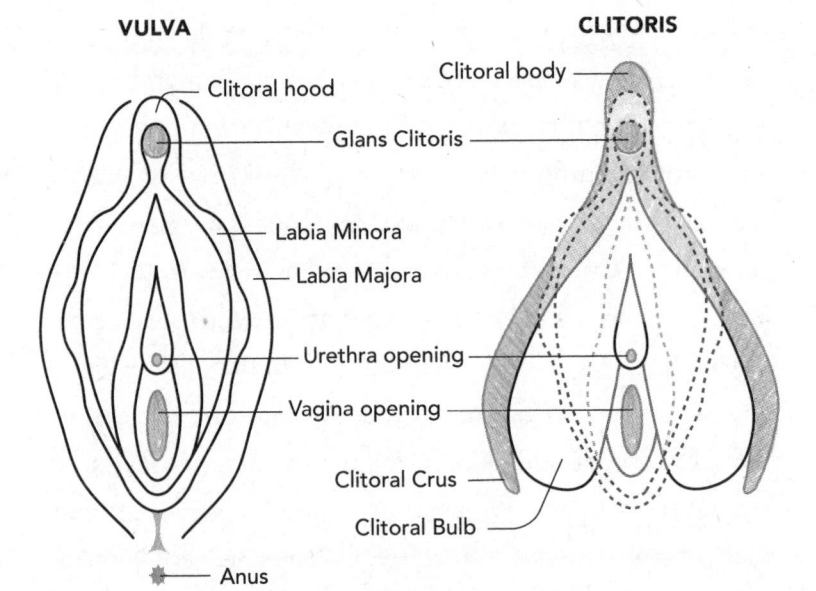

Vulva (left) and clitoris (right).

The glans clitoris is incredibly dense with nerve endings that come into play during sexual arousal and orgasm. The glans has so many nerve endings that for some women, it is too sensitive to touch directly. The glans does swell during sexual arousal, but it has less erectile tissue than other parts of the clitoris. Erectile tissue, with its many blood vessels, smooth muscle, and connective tissue cells, has a spongy texture and becomes erect or swollen when extra blood flows to it. The subterranean body, crura, and bulbs of the clitoris are composed mainly of erectile tissue, containing fewer nerve endings than the glans.

With this anatomy in mind, let's now discuss what happens

to the clitoris and other parts of the body during an orgasm. At a physiological level, orgasms are a sensory-motor reflex. Sweating when it is hot outside or quickly moving your hand away from a burning object are examples of sensory-motor reflexes, which happen without any conscious input from your brain. The textbook sensory-motor reflex is the knee jerk. Your doctor's tap on your knee activates sensory nerves in your leg, which sense information from the outside world and carry it to your central nervous system—in this case, just the spinal cord. In the spinal cord, the sensory information is relayed to motor nerves, which are the nerves in your body that tell your muscles to take action. The knee jerk happens when the activated motor nerves cause the muscles in your leg to contract.

While an orgasm is much more complicated, involving many more nerves, muscles, and your brain (not just the spinal cord), it is also a reflex. The sensory input is received by nerve endings in the clitoris, and the motor output (the reflex) is the involuntary contraction of muscles in the pelvic region, including muscles in the pelvic floor and in the vagina and uterus. The brain is also involved as sensory information gets relayed from the clitoris and translated to muscle contractions during orgasm. Many parts of the brain become activated, and many hormones and neurotransmitters are released. Dopamine, serotonin, and endorphins, all known to make you feel good during certain activities, are released in the brain leading up to and during an orgasm. There is also a surge of a few hormones—prolactin, oxytocin, and luteinizing hormone (LH). Prolactin is known for its role during lactation (we'll talk about it in a later chapter), but it is an ancient hormone involved in many processes including ovulation, embryo implantation, and labor in some species. Oxytocin, a hormone that gets released from the brain during social interactions (it's been given cutesy names like the "cuddle chemical" and the "hug hormone"), also does many things in the body, including causing the muscle contractions during labor, lactation, and ovulation. And remember from the last chapter that LH plays a critical role during ovulation.

Another difference between the knee jerk and orgasm is that the orgasmic reflex lasts longer, with muscle contractions of the pelvic floor, vagina, and uterus continuing for up to a minute. It also takes much longer to achieve an orgasm after the sensory nerves first become activated in the clitoris. On average, it takes about four minutes of sustained clitoral stimulation to reach orgasm by masturbation, which is the fastest and most reliable way for women to achieve an orgasm. William Masters and Virginia Johnson, who pioneered research on the human sexual response starting in the 1950s, defined four stages of orgasm (and this definition is still used by many today): excitement, plateau, orgasm, and resolution. During the excitement stage, stimulation of sensory nerves in the clitoris causes an involuntary increase of blood flow to the genital region. The small arteries in the clitoris and surrounding areas dilate, allowing more blood to flow to the region and resulting in engorgement of these tissues because they are spongy. Independent of the muscle contractions during orgasm, the dilation of blood vessels in the excitement stage is a reflex on its own, and in men, it results in an erection. The plateau stage is a sustained and heightened level of excitement. If and when a threshold is reached, the orgasm stage occurs, which involves up to a minute of involuntary contractions of the pelvic floor and the feeling of intense pleasure. The contractions start pushing out the blood that had pooled in the genital tissues during excitement and plateau, and the resolution stage is when the genital tissues have become completely drained of blood and return to a pre-excitement state. This is a working model; some have added "desire" to the beginning of the sequence, and some have noted that multiple orgasms are not well described by the model. In response to the latter point, Masters and Johnson reasoned that after the orgasm stage, a woman (but not a man) can return immediately to the plateau stage, have another orgasm, repeat this partial sequence multiple times, and then slowly return to the pre-excitement stage after the last orgasm.

Adaptation or accident?

Now that we have a better physiological understanding of what an orgasm is, let's look at the debate on why women have orgasms in the first place. The contemporary conversation started when anthropologist Donald Symons wrote a book in

1979 called *The Evolution of Human Sexuality*, a chapter of which focused specifically on the female orgasm. Symons proposed that the female orgasm is simply a by-product of male orgasm. In other words, the female orgasm is a spandrel, to use Stephen Jay Gould's vocabulary describing the phenomenon of evolutionary by-products. Gould himself got involved in the debate in 1987, strongly backing Symons's by-product hypothesis.

There are two foundational pillars of Symons's by-product account. The first is the well-agreed-upon fact that male orgasm is under strong natural selection. Contractions of the pelvic floor muscles and prostate gland during male orgasm push sperm into the urethra and out of the penis, and these contractions are critical for a male's reproductive success. Without ejaculation, he's unlikely to pass his genes to the next generation.

The other foundation of the by-product account is the shared development of young female and male embryos. The penis and clitoris develop from the same embryonic nub of tissues during development, so the fully formed structures are composed of the same erectile, muscle, and nervous tissue platform that allow for an orgasm. The clitoris and penis both have a glans, roots, and bulbs, even if they are organized a little differently in the adult. Females have orgasms, then, because of shared development and strong selective pressure for them in males. They are a lucky accident and are inconsistently experienced during vaginal intercourse because they have no evolutionary purpose in women.

Symons's hypothesis was met with skepticism, attack, and the introduction of a suite of adaptive explanations for the female orgasm in the 1980s and '90s. To many biologists at the time (and still today), the notion that a trait as complex as the female orgasm might exist by accident is outrageous. Symons's by-product theory has also been viewed as antifeminist, tethering female sexual response to that of a male, denying the importance and uniqueness of female sexuality and pleasure (more on this later). But to be fair, many of the adaptive accounts, briefly reviewed below, are also conspicuously male-centric.

The first set of adaptive explanations in the 1980s proposed that the female orgasm evolved by natural selection in order to promote "pair-bonds" between females and males. The pair-bond isn't a lifelong monogamous marriage, exactly, but it involves an extended period of time in which both parents raise their child(ren) together. Over ten pair-bond explanations of the female orgasm have been proposed, most of which are based on two assumptions. First, the pair-bond that evolved between females and males in our human ancestors was an adaptation—those individuals who engaged in pair-bonds achieved more reproductive success than those who didn't. Second, the female orgasm evolved to strengthen the pair-bond. For example, a woman who achieves an orgasm during intercourse is more likely to want to have sex again with her partner, will be more motivated to care for her partner, or will choose to stay bonded to the more attentive lover over a less attentive one, because a better lover is a better parent. The pair-bond hypotheses are all variations on this same theme. By the 1990s, though, pair-bond accounts of female orgasm had largely been discredited because it became clear that many animals that form pair-bonds, including humans, often mate with individuals outside the bond.

Other adaptive accounts have also been proposed, suggesting alternative functions of the female orgasm that have nothing to do with strengthening the pair-bond. In one account, the female orgasm, which allegedly precedes male orgasm during sexual intercourse, evolved because the contractions of the vagina help males reach sexual climax. In another, the female orgasm evolved to signal to the male that the female is sexually satisfied and thus disinclined to seek sex elsewhere, which is adaptive because it supposedly protects her from being harmed by a jealous partner. As one detractor has argued, this seems like a better hypothesis for why women evolved to fake orgasms (like Sally).

The most recent set of adaptive explanations is based on the idea that the muscular contractions during female orgasm help "upsuck" sperm into a woman's reproductive tract. The upsuck of sperm is hypothesized to increase her chance of fertilization. The linchpin of the upsuck accounts is that a female uses the orgasmic upsuck as a way to control whose sperm fertilize her eggs. She will only have an orgasm with a "superior" male, whose genes she wants for her children, who by the way may not be the guy with whom she's pair-bonded.

The upsuck hypotheses emerged to explain why females have sex with males outside the pair-bond and why they (supposedly) have orgasms with these males and not their partners. According to the upsuck accounts, orgasms allow a woman to parent with the good dad who provides lots of resources but have children with the sexier, genetically superior male.

In 2005, philosopher of science Elisabeth Lloyd systematically savaged each adaptive account of female orgasm in her book *The Case of the Female Orgasm*. As Lloyd outlines in her book, many accounts—in particular the early pair-bond accounts—completely ignore or misinterpret the literature on female sexual response that shows that orgasm during vaginal sex is highly variable, both between women and in the same woman. In other words, a high percentage of women never have an orgasm during vaginal sex, and of those who do, they don't climax every time. This variability of female orgasm during reproductive sex weakens the argument that orgasms enhance a woman's reproductive success—if the female orgasm was an adaptation, the trait would be much more consistently displayed as it is in males. Moreover, many studies have failed to find evidence that orgasm has any effect on fertility or reproductive success in women, which is a foundational assumption of all the adaptive accounts. Since Lloyd's book was published, a large study with solid statistics again showed no correlation between orgasm rate and the number of children a woman has.

Lloyd saves her most ravaging critique for the upsuck

hypotheses, which at the time her book was published were still being seriously considered. In addition to highlighting many of the questionable assumptions underlying the upsuck accounts, including that women seek out males with superior genes (who are outside their pair-bonds), that they can detect these superior males, and that they more easily achieve orgasm with these superior males, Lloyd picks apart the analytical methods used in these accounts. First, the original observation of the upsuck phenomenon was in one woman on two instances—this is a tiny sample size. In addition, upsuck of sperm wasn't even measured in this woman—it was a difference in pressure between inside the uterus and outside that was observed during orgasm. Many other studies that specifically tested for upsuck of fluid during orgasm failed to find an upsuck effect. Nevertheless, a series of later studies were based on the assumption of upsuck during orgasm, even though there is no evidence to support it. Moreover, Lloyd shows that the authors of these later studies, who have argued that upsuck is an adaptation that allows females to choose sperm from genetically superior males, inappropriately manipulated their data (which included survey responses and physical data like sperm counts and volumes of vaginal fluid from women after sex). According to Lloyd, they also used the wrong statistical tests to analyze their data, rendering their results worthless. So not only did these authors base their hypotheses on the questionable idea of upsuck, but they also massaged their data to support their claims. Her pointed criticism extends to the reviewers and editors of the journals who allowed those flawed papers to be published.

One of the main theses in Lloyd's book is that a blind faith in adaptationism—the conviction that all biologically "important" traits must be the direct product of natural selection—has driven some in the field to extreme measures, including ignoring relevant literature and inappropriately manipulating data to get the preferred answer. Not surprisingly, Lloyd supports Symons's by-product account of the female orgasm, which she argues is the only account thus far that is not based on flawed assumptions and that most accurately describes the trait of orgasm as it is experienced by women today. Lloyd also addresses the feminist objection to Symons's account, which is essentially that a by-product explanation of the female orgasm denies and devalues its importance in woman. Lloyd argues that a by-product origin of female orgasm does not dictate our cultural views toward it—a trait can be culturally important without being an adaptation. She reasons that our ability to play music, for example, is not an adaptation, but it is still a valued and celebrated human trait.

You might think Lloyd's book closed the case of the female orgasm, but it did not: The debate continues. Some in the adaptation camp continue to insist that the lack, as of yet, of a well-supported adaptive explanation does not mean it does not exist. In addition, voices new to the scene propose a hypothesis that doesn't quite fit in either camp. As you'll learn next, this explanation opens up a new—or perhaps I should say ancient— look at the evolutionary relevance of the female orgasm.

Orgasmic origins–triggering ovulation in early mammals?

The voices that recently joined the scientific conversation on female orgasm are ones I know well. As I mentioned earlier, Günter Wagner was my thesis advisor and Mihaela Pavlicev is a colleague. In their take on the female orgasm, they remind us that the functions of biological traits don't necessarily remain static over time. They can emerge to fulfill one function but then change or lose that function later (remember our discussion of bird feathers from the introduction). Pavlicev and Wagner's complaint about the debate on female orgasm is that it has focused on the evolutionary function of orgasm (or lack thereof) in human women today, which overlooks its original purpose in our mammalian ancestors.

I was also guilty of a human-centric view of biology when I started my thesis work with Wagner, insistent on studying some aspect of human evolution because humans were more interesting to me than chickens, opossums, and the other critters we studied in the lab. But as he repeatedly reminded me while I was training with him (sometimes in total exasperation with my predoctoral naivete), most human traits are just modified versions of ancient animal traits. So to understand how and why the female orgasm (or any other trait) exists in humans, we must look at the broader evolutionary context.

Pavlicev and Wagner noticed that what happens hormonally during an orgasm in women—a release of prolactin, oxytocin, and, according to one study, LH—is similar to the hormonal surge that occurs in some female mammals during copulation, which is the act of inserting the penis into the vagina for sperm transfer. Rabbits, cats, ferrets, and squirrels are some of the mammals that are induced to ovulate during copulation. Ovulation in these mammals is triggered by physical stimulation of the vagina (and Pavlicev and Wagner think the clitoris, a structure all female mammals have) by the penis. The sensory information from the vagina/clitoris during copulation is carried to the brain, triggering the release of hormones in the brain that cause the release of egg(s) from the ovary. This egg-releasing strategy makes intuitive sense—only release an egg when there's sperm around to fertilize it. That's not, of course, what happens in humans—we ovulate "spontaneously" around the same time every cycle, regardless of whether or when we've had intercourse. Spontaneous ovulation does not require any external signals, only the hormones made in the female's body itself.

To understand how orgasm and ovulation might be connected, Pavlicev and Wagner take us on a tour of ovulation in different species in the animal kingdom. The axis of organs and hormones that control ovulation is remarkably similar in distantly related species, from fish to humans. We talked about this axis in previous chapters: The brain makes and releases hormones (GnRH, FSH, and LH) that stimulate the ovaries to make estrogen and to mature and release eggs. What varies across animals is the trigger at the top of the cascade, the signals that tell the brain to start releasing its ovary-stimulating hormones. In most fish species and many land-dwelling animals as well, the cascade is triggered or influenced by environmental cues like temperature or the amount of daylight or moonlight. Horses, for example, only ovulate during late spring and summer months when there is more daylight. The precise environmental cues vary according to what conditions are most important for successful reproduction in each species. The time of year is critical for many. Several species have evolved to produce babies during the season(s) when food will be abundant, so mothers only develop and ovulate eggs at the times of year that will allow that to happen.

Another point is that many of the animals that use environmental cues to trigger ovulation undergo *external* fertilization, which is the fertilization of eggs outside the female's body. In most fish and amphibian species, females release their eggs into the water right after ovulation, males release their sperm to fertilize the eggs, and the fertilized eggs develop from there. In contrast, amniotes (the group that includes mammals, reptiles, and birds) use the strategy of *internal* fertilization, in which the male deposits sperm directly into the female's reproductive tract. External fertilization leaves eggs vulnerable to predation and drying out, making the environmental conditions at the time of ovulation and egg release especially important. This is another reason why many animals time ovulation (and egg release from the body in external fertilizers) to occur when environmental conditions are most conducive to successful reproduction.

In their research, Pavlicev and Wagner focus on the triggers of ovulation in mammals. We know that environmentally triggered ovulation is extremely ancient, since we observe it across all types of animals, from fish to mammals. Using evolutionary analytical tools, Pavlicev and Wagner have shown that in the first mammals, the trigger had likely changed to the stimulation of the vagina/clitoris during copulation (probably in addition to some seasonal cues). Many mammals today still use copulation as a trigger. Rabbits, cats, and other species that are solitary and/or have large home ranges are induced to ovulate during the act of copulation. This strategy works well in species in which the chance of finding a mate puts limits on reproduction. In these mammals, sensory information taken in by the vagina and clitoris during copulation is the cue that tells the brain to release the hormones that will induce ovulation.

But in some groups of mammals, like primates, bats, elephants, and some rodents, the trigger for ovulation changed yet again, and spontaneous ovulation evolved. This is the strategy taken in our own species—there is no external cue (like the length of day or the presence of a male) that tells our ovaries to release eggs. The brain-ovary axis controls ovulation all on its own in a sophisticated feedback loop that occurs in cycles.

A remarkable and critical connection that Pavlicev and Wagner make is between the type of ovulation in mammals—spontaneous or by copulation—and the location of the clitoris relative to the vagina. They show that in ovulators by copulation, the glans clitoris—the part that is full of nerve endings—is *inside* the vagina or on the border of it. According to their model, this location ensures that the brain will receive the sensory signal to ovulate during the act of copulation. In contrast, in spontaneous ovulators, the glans clitoris has drifted away from the vaginal canal and is often located a relatively long distance away (in humans, it is located on the other side of the urethral opening). Because spontaneous ovulators don't use sensory signals during copulation anymore to trigger ovulation, the ovulatory function of the clitoris became superfluous. In spontaneous ovulators like humans, then, the clitoris was free to change without any constraint on location or function. In humans, it has been established that the long distance between the glans clitoris and vaginal opening is the reason many women have a difficult time achieving orgasm during vaginal intercourse. But the capacity to have an orgasm, in particular by masturbation, remains with us as a relic of its ovulatory role in our mammalian ancestors. Some have even argued that this freedom from evolutionary constraints has allowed the orgasm in women to become more pleasurable than it is in other species.

Wagner and Pavlicev have provided further support for their hypothesis by doing experiments on rabbits, one of the species that ovulates during copulation. They treated these rabbits with the antidepressant fluoxetine, which is known to inhibit orgasms in women. As they predicted, fluoxetine also inhibits ovulation in rabbits, supporting their hypothesized connection between ovulation and orgasm.

To wrap it up, Pavlicev and Wagner have developed a new explanation for the origin of the female orgasm. They argue that it had a critical reproductive function—inducing ovulation—in early female mammals (and in many mammals still today). It was adaptive in the past, but in humans and some other species, it lost its ovulatory function, which explains some of its puzzling qualities in the modern woman.

An even more ancient take on the female orgasm

For all this talk about orgasms in other mammals, how do other animals experience orgasms? While many of us know what an orgasm feels like and looks like when another person is having one (or faking one!), what is the evidence that other animals experience orgasms the way that we do or at all?

While it is impossible to establish how a rabbit or monkey feels during sex since they can't tell us about it, many studies have shown that animals do exhibit the sensory-motor orgasmic reflex and that it may even be pleasurable to them. A number of observational and experimental studies have been done in nonhuman primates. Both male and female stumptail macaques exhibit similar behaviors at the end of some sexual encounters: The muscles of their bodies spasm, they make a round-mouthed facial expression, and they cry out in a distinct way. While these behaviors in males are tied to ejaculation during copulation, the behaviors have only been observed in females when they mount and rub their genitals against other females. Female bonobos also frequently rub their genitals against each other, and some have been observed to have spasms of the pelvic floor muscles at the end of these encounters. Female Japanese macaques have been observed to have contractions of the pelvic floor muscles during masturbation, which is an activity performed by *many* animals (both male and female), including turtles, birds, walruses, and squirrels. Back when research funding flowed more freely than it does today, orgasms were studied in many primate species by a researcher manually stimulating a female's clitoris. All the monkeys and apes that were studied in this way appear to have the ability to reach a sexual climax. While the duration of masturbation and copulation is much shorter in nonhuman primates, on the order of seconds (which perhaps is an indication of how less pleasurable sex is to them compared to humans), a large body of work suggests that females and males in many primate species can experience orgasms.

Looking outside the primates, when the genitals of male and female rats are stimulated during copulation, both sexes display many of the physical and behavioral changes that humans do during orgasm. The muscles of the pelvic floor and reproductive tract contract, many of the same neurotransmitters and hormones are released, and their behavior in the short and long term suggests that orgasm is rewarding and something they want to experience again. Moving even farther afield, the male buffalo weaver bird has an unusual phallic structure that appears to be purely sensory (it doesn't carry sperm or urine). He rubs the structure against the female's hind end during sexual encounters, and as the male ejaculates from his cloaca—the shared opening for waste and reproductive products in many animals—his wings quiver, his body shakes, his legs spasm, and his feet clench. In experiments in which buffalo weaver bird males copulated with a model female (imagine an inflatable bird sex doll), the only time they exhibited these orgasmic behaviors was when ejaculate was found in the model's cloaca.

Even fish species, most of which do not have copulatory organs because fertilization takes place externally, exhibit a reflex that is reminiscent of an orgasm. After eggs and sperm are fully developed and ready for fertilization, both males and females in many fish species release their sex cells into the water in a process called spawning. Spawning requires a lengthy series of muscular contractions of the pelvic region and reproductive tract, and other parts of the body often shake or quiver too. These contractions are referred to as "shuddering" in the fish reproduction literature. Fish shuddering is what got me thinking about how old some components of the orgasm might actually be.

What I'm about to propose in the next few paragraphs is entirely speculative, but I'd like to see Wagner and Pavlicev's hypothesis and raise it a few hundred million years. One of the complaints about their hypothesis (by Elisabeth Lloyd among others) is that they focus only on one aspect of orgasm—the hormonal surges in the brain. But orgasms are a complex sensory-motor reflex that starts with sensory information at the clitoris and ends in the contractions of multiple muscles in the pelvic region and reproductive tract. Wagner and Pavlicev don't directly address those muscular contractions, which are a central component of the orgasmic reflex.

How far back in time do we need to go to understand the origin of the reflex contractions that are a defining feature of orgasms? I propose going back over 500 million years and looking at all vertebrates, which include fish, amphibians, reptiles, birds, and mammals. The similarity of the ovulation cycle in fish and humans that I mentioned above is not a coincidence—we share a common ancestor with all vertebrate species on the planet today, which lived in the oceans and had some version of the brain-ovary reproduction axis that controls ovulation. We talked about fish ovulation earlier; depending on the species, it happens when the length of daylight or perhaps the water temperature is just right, which tells the fish brain to release the hormones that trigger egg development and ovulation from the ovary. We also discussed external fertilization in fish—ovulation in many fish species is quickly succeeded by spawning. These two processes, ovulation and spawning, are closely associated and influenced by some of the same hormones. Fish farmers treat fish that won't ovulate or spawn with mushed-up pituitary glands from fish brains, which contain the hormones we've been discussing—LH, prolactin, and oxytocin. Might some components of the orgasm trace back to ovulation *and* spawning in our fish ancestors?

This ancient reflex must have been quite different from what we identify as an orgasm today. Most fish have no need for a clitoris, penis, or any other localized sensory organ for reproduction, since they don't have intercourse. Instead, environmental cues (and social cues in some species) tell the brain to trigger ovulation and spawning in females and spermiation (the release of mature sperm into the male reproductive ducts) and spawning in males. Environmental/social cues are a different kind of sensory information that doesn't require a localized organ or structure in the pelvic region to integrate that information into the body.

But when internal fertilization evolved in amniotes, there was need for a penis or at least a strategy for males to deposit sperm into the female's reproductive tract. Many reptiles have a penis for sperm transfer. Some birds (like ducks) have a penis, but many birds just push their hind ends together in what is called a "cloacal kiss." All mammals have a penis, although the structure has been modified extensively in different species. Although there is tremendous diversity of male external genitalia across amniotes, recent research suggests that the embryonic nub of tissue that develops into a penis or clitoris in humans is quite ancient. All amniotes have that embryonic nub (even if they don't develop a copulatory organ from it), suggesting that the penis—and the clitoris—originated once in early amniote history over 300 million years ago. Then it was modified in individual lineages and species according to the specific needs and constraints in each group.

As internal fertilization and the penis first evolved in our amniote ancestors, you can imagine that it was now useful for males to release sperm into the female reproductive tract during the act of copulation (instead of or in addition to environmental cues). The muscular reflex causing sperm release is an ancient fish trait—shuddering—as we just discussed. But as the penis first evolved, it would have been advantageous for the muscular reflex to be hooked up to the penis, requiring a rewiring of the reflex to the new sensory/copulatory organ. This rewiring is evident in our amniote relatives, such as some species of snakes and lizards, whose penises can be manually stimulated to ejaculate sperm for assisted reproduction in zoos and conservation centers.

As the penis and new wiring were evolving in males, similar changes would have been occurring in females. Since the clitoris and penis are derived from the same embryonic nub of tissue in all amniotes, the neuro-genital changes happening in males would have been occurring simultaneously in females. The reflex of ovulation and spawning already existed. Now they would have been wired up to the new sensory organ—the clitoris. Some of our amniote relatives today, like certain species of turtles and snakes, ovulate during the act of copulation, suggesting that clitoris-mediated, copulation-induced ovulation is even older than Pavlicev and Wagner have suggested.

But additional reproductive changes would have been necessary in our female amniote ancestors with internal fertilization. I hypothesize that the contractions of female orgasm today are a relic of their ovulation/spawning function in our fish ancestors, but our amniote ancestors would have needed to delay the completion of "spawning" long enough for copulation and fertilization to take place and long enough for the preparation of fertilized eggs before laying (most reptiles/birds) or the extended development of babies before birth (most mammals and some reptiles). The muscle contractions during egg laying in reptiles and birth in mammals are influenced by some of the same hormones we've been discussing, in particular oxytocin and possibly also prolactin. When I reached forty-one weeks of pregnancy with two of my children, I was induced into labor with Pitocin, which is a synthetic version of oxytocin.

Most relevant to how we experience orgasms today, as these neuro-genital changes were taking place in our amniote ancestors, there was also likely additional pressure for the reflex to feel rewarding. Behaviors that are critical to an individual's reproductive success, like eating and copulating, activate neural circuits in the brain that make an individual feel pleasure and motivate them to engage in the behaviors again. The phenomenon of pleasure is one of evolution's sneakiest tricks, motivating individuals to engage in behaviors that are necessary for reproductive success. These neural circuits of pleasure and reward are ancient, found in all vertebrates. So the evolution of the modified sensory-motor orgasmic reflex in amniotes likely also involved activating pleasure and reward centers of the brain that encouraged individuals to engage in copulation.

At this point, I've connected back to Wagner and Pavlicev's hypothesis. In early mammals, the function of the clitoris and orgasm was to integrate sensory information during copulation as a trigger for ovulation. But in many species, like humans and other primates, the ovulatory function of the clitoris was subsequently lost when females took complete control over their ovulatory cycles.

Putting it all together, what I'm proposing (and again, much of this is speculative and requires further testing, but I'm a researcher and that's what we do) is that the female orgasm has a long and convoluted history. In our fish ancestors hundreds of millions of years ago, there existed an environmentally triggered reflex to ovulate and release eggs into the water. This reflex involved a surge of hormones in the brain, ovulation, and muscular contractions of the reproductive tract and pelvic region to expel eggs from the body. When internal fertilization and a penis/clitoris evolved in our land-dwelling amniote ancestors, the ancient reflex was modified in females (there was now a delayed release of fertilized eggs/babies from the body), and it got wired up to the new sensory organ and to pleasure/reward centers in the brain. In some mammals today, the ovulatory function of the clitoris and orgasm is preserved, but in many spontaneous ovulators, such as humans, this function has been lost. All that is left are traces of its earlier functions (the hormonal surge and the pelvic/uterine/vaginal contractions) and of course the capacity to make us feel great.

It is this protracted evolutionary history that makes the female orgasm such an enigma. It is an adaptation, a by-product, a modified trait, and a lost trait all in one. The orgasm's recent change in our primate ancestors—the loss of its ovulatory function—makes the female orgasm both frustrating and fabulous for women: frustrating because the drifting away of the clitoris from the vagina makes it difficult for some women to achieve orgasm during vaginal sex, and fabulous because the clitoris was liberated, having no function in a woman other than bringing her pleasure. As some have suggested and as we'll discuss in the next chapter, the liberated clitoris continued to change in the human lineage, evolving the capacity to bring even greater pleasure to women than it appears to bring to other female mammals. If it were possible for a woman to sit down in a New York deli with some lady mammals, who don't appear to experience sexual pleasure the way that women in our species do, I can imagine one of them looking over and saying, "I'll have what she's having!"

CHAPTER 5

What's Love Got to Do with It?

No matter how cynical I become with age, I think I'll always fall for a good love story. The variations are endless—there are the romances between the poor woman and the rich man (*Pride and Prejudice, Pretty Woman*), the rich woman and the poor man (*Titanic, The Notebook*), old friends (*Emma, When Harry Met Sally...*), members of the same sex (*Carol, Brokeback Mountain*), members of different species (*Avatar, Twilight*). I'll admit that my favorite movie romances are the ones with dancing, no matter how wretched the acting is (*Dirty Dancing*, also a rich gal/poor guy romance). Stories about falling in love never seem to get old and may go as far back as our ability to tell them.

As evidenced by the number of songs, poems, novels, movies, and movie remakes about love, humans clearly have a fascination with it. Why do we fall in love? This isn't a topic you'd think evolutionary biologists have much to say about, a topic more appropriate for the realms of music and multiplexes.

But if we phrase the question slightly differently—why are we choosy about our mates?—there is in fact an extensive body of research on the subject. Just like stories about love, the scientific discussion on mate choice is long-standing, the first ideas presented by Darwin in his book *On the Origin of Species* in 1859 and elaborated on in *The Descent of Man* in 1871.

What did Darwin have to say about mate choice? As we discussed in earlier chapters, Darwin's adaptation by natural selection explains the evolution of traits that increase the odds of surviving and having more kids. The fastest land animal in North America—the pronghorn antelope, which can run close to sixty miles per hour—likely evolved its speed to avoid predators that roamed the area over ten thousand years ago, such as the American cheetah. But Darwin famously struggled with some traits, like the peacock's flamboyant feathers, that would seem to do the opposite—its conspicuous and cumbersome train should attract predators and make it more difficult to survive. Darwin's answer to that puzzle was his idea of sexual selection, which explains the evolution of traits that enhance an individual's ability to get mates for reproduction, even if they come at a cost of reduced survival. He described two categories of traits that evolve by sexual selection: traits in one sex—usually male—like large size and body weaponry, that enable males to compete with each other for access to females; and traits in one sex—also usually male—that appeal to female perceptions of beauty, their "taste for the beautiful" as Darwin wrote. Sexual selection by mate choice (the second category) is more complex, interesting, and controversial than sexual selection by male-male competition (the first category), because it involves at least two traits in two different individuals—the beautiful trait in males and the *preference* for the beautiful trait by females. Darwin surmised that the evolutionary back and forth between preferences and beautiful traits over time led to the highly elaborate, stunning, and unique traits we observe today across the animal kingdom.

Since Darwin, biologists have gained some fascinating insights into the who, what, when, where, and why of mate choice. Humans aren't the only animals who make choices about with whom they couple up, and most of the work on mate choice has focused on nonhumans. In many species in the animal world, including humans, researchers have asked the following questions: Who is doing the choosing: female or male, both, or neither? What traits do choosers consider attractive? When and where are choices about mates made? And most controversial of all our W questions, why are individuals choosy in the first place? These are the questions we'll tackle in this chapter, always with an eye on our target: What's mate choice got to do with the evolution of our biology? Tina Turner might not agree, but a large body of work shows that it has a lot to do with it!

The who

I've lived in the same house for three years in Northern California with my family. The year we moved in, we noticed small birds nesting in one of the lamp fixtures outside our garage. We eagerly watched the transformation from a sparse nest to a comfy-looking one, from an empty nest to one with four white eggs with dark speckles, from a nest with quiet, helpless hatchlings to one with boisterous, restless fledglings, and finally to an empty nest again. During our second year in the house, a pair of birds decided to nest in a different spot, on top of the vent for a gas fireplace that we never use. It wasn't quite as exciting the second time around, in particular for me—an admitted clean freak—because this spot is located above some outdoor furniture that became covered in droppings. This year, we've got nests (and poop!) everywhere. The message seems to have gotten out that our yard is a suitable place to live. We even watched one bird try repeatedly and in vain to nest in a location above our back door that simply can't hold a nest. For a week, we were reminded of the determined bird's efforts every time we went outside and had to step over the thin twigs and other materials that kept falling to the ground.

Based on the appearance of the eggs, nests, and birds themselves, I've identified our backyard friends as house finches. House finches are native to western North America, but in the 1940s, pet shop owners in New York City released hundreds of them to avoid prosecution for illegally selling them, and now the birds can be found across the entire United States and some parts of Canada. House finches are gregarious, nonterritorial, nonmigratory birds—in other words, they are social, mild-mannered birds that stay close to home all year long. In between breeding seasons, they flock and forage together in large groups of up to hundreds of individuals. In my neighborhood, I see them sitting together on power lines and fences and foraging together in an open nature preserve where I like to hike. But during the spring and summer breeding season, they pair up in couples. House finches are notably flexible about where they start their families. They nest in a variety of trees but are also happy to do it in crevices around buildings and in other human-made structures like streetlamps and hanging planters. Female and male house finches look different from each other. Both sexes are a drab, streaky brown, but the males are decorated with a splattering of red on their heads and throats. Female and male house finches also sound different from each other. Both sexes talk, but only the male sings. The chorus of male house finches fills my backyard, a spring and summer concert of elaborate buzzes, chirps, and twitters. Differences in appearance (such as red versus brown head) and behavior (such as elaborate versus simple songs) between males and females in a species are often a sign that sexual selection has taken place. The house finch, then, is a good species close to home to start our discussion of mate choice and sexual selection.

Of all the potential mates in a house finch's large social group, how do individuals decide on one mate at the beginning of the breeding season? Is it mutual love at first sight, like between Tony and Maria at the gym dance in *West Side*

Story? Or does one individual have to work tirelessly to woo a skeptical other, like Adam Sandler's character courting Drew Barrymore's in *50 First Dates*? Or is the process of getting a mate more sinister, with one party using violence and intimidation to coerce the other into the "relationship"?

To answer this question, especially in birds, one of the first things to look for is any sign of showing off. The show-offs in the animal world are usually trying to court picky potential mates. A dorky Adam Sandler shows off by building waffle houses and enlisting the help of an adorable penguin, but male house finches do it with their bright red feathers, their elaborate songs, and their displays during the breeding season—fluttering upward while singing and then gliding down quickly. At first glance, then, it appears that the male house finch is the courter trying to convince skeptical females to choose him instead of the other guys. Many studies on house finches, both in captivity and in nature, have shown that females prefer brighter red males over duller red or orange ones, and they prefer the males who sing longer, more complex songs.

The tagline in the study of mate choice has been that females are the pickier sex because they invest more in making and raising offspring than males. We've discussed this in previous chapters—females produce larger, more energy-rich sex cells (eggs) than males (sperm). In many species, they nourish and incubate their developing fetuses and nurse, feed, and protect their young offspring. What limits reproductive success in females is the number of offspring they can

of satin bowerbirds, which build upright sex dens with twigs and then fastidiously embellish them with blue items from their surroundings. In these species, females shop around until they find what they want, which is often the most extreme display, whether it's the brightest, longest, loudest, or most ornamented. In one classic experiment on African widowbirds, researchers cut the tails off some males, glued them onto others, and showed that the males with artificially lengthened tails were more successful with females, even though their extreme tail lengths are never observed in natural populations. Once a female has mated with the sexiest male she can find, he leaves, and she's saddled with all the baby work. But presumably she reaps the evolutionary benefits of her good choice when her offspring are more successful than those of other females who aren't as choosy (we'll explore this later).

Of course, it would be too good to be true if mate choice were this simple to explain or generalize. There are many mammalian species in which females do all the work of raising offspring, but their freedom of choice is limited. In species that reproduce in harems, one very large male controls the sex life of a number of smaller females. Male elephant seals and gorillas are prime examples—they are enormous relative to the females, and they use their size to keep other males and females in check. For years, it was thought that the only freedom a female elephant seal had was when choosing the dominant male of her harem, called the beachmaster, over smaller males who try to sneak copulations with her. When sneaky males approach, female elephant seals have been observed making loud noises to alert the beachmaster of the usurper. However, it appears females in one colony on Marion Island (between Africa and Antarctica) have more choice than previously thought, sometimes opting to mate at sea with a male of their choosing rather than on land with a beachmaster.

Female langur monkeys don't have the choice of sex at sea, but they often choose to mate with the sneaky males. Primatologist Sarah Blaffer Hrdy has argued that promiscuous behavior in female langur monkeys evolved for a very specific reason—to prevent future dominant males from killing their babies, a phenomenon known as infanticide. Infanticide occurs in many primate societies when a dominant male is overthrown by a newcomer. The victorious newcomer proceeds to kill the current young offspring in the harem in order to more quickly impregnate females who wouldn't otherwise ovulate while they are still breastfeeding their children with the overthrown male. The preference of female langur monkeys to mate with sneaky males, however unimpressive as they might be, likely evolved to confuse paternity, since a new dominant male is less likely to kill a female's offspring if he has ever mated with her in the past. If there's any possibility that those children are his, he will not kill them. Thus, like elephant seals, female langur monkeys do appear to exert some choice, although I'm thankful I'm not faced with those options.

In elephant seals and langur monkeys, there is no doubt that sexual selection has taken place, resulting in a remarkable size difference between the sexes and male body weaponry like sharp canine teeth. But it appears that sexual selection processes other than mate choice have dominated. Male-male competition, as Darwin outlined, was certainly one of them, but there are additional mechanisms that Darwin did not describe in detail that were also at play. In many species with large, weaponized males, these males use their size and armament not only to compete with each other but also to force females to copulate and thwart their freedom of choice, which is known as sexual coercion. Like infanticide, sexual coercion is a manifestation of sexual conflict, in which the evolutionary interests of one sex are advanced at the expense of the other. Strategies like sexual coercion and infanticide by males are often countered by defensive strategies by females, whether it's simply running away or mating with multiple sneaky males in order to confuse paternity.

Just like the back-and-forth evolution of preferences and beautiful traits in some species, the escalation of coercive and defensive strategies in other species, and sometimes even in the same species, has resulted in some extreme traits and behaviors in the animal world. In mallard ducks, sexual selection by mate choice is evidenced by lustrous green-headed males and dull brown females, but in addition, there has been a sexual arms race in the species. Males are coercive, forcing unwilling females to copulate. They have additionally evolved a long counterclockwise-oriented corkscrew-shaped penis, and females a *clockwise*-oriented corkscrew-shaped vagina, in a conflict over the female's right to choose her partner's sperm. The females are not successful at avoiding the forced copulations, but they evolved an unusual vaginal shape to prevent sperm from unwanted males from fertilizing their eggs. Female ducks are not alone. Bottlenose dolphin vaginas have intricate pleats and folds that may also provide a female some freedom of choice, not over with whom to mate (male dolphins can be extremely aggressive) but over whose sperm she'll allow to fertilize her eggs. Females may position themselves during copulation and contract or relax their vaginal muscles to either steer unwanted sperm toward one of the folds, which are dead ends, or wanted sperm toward their eggs. The evidence of sexual conflict in these species is written into the shape of their genital structures, the bewildering qualities of which only make sense in light of the conflict.

I've been focused on females who provide most of the childcare, but there are some striking role reversals in the animal world, in which males are the sole childcare providers and females have evolved to impress. In birds called phalaropes, which are native to Iceland, the females are larger and more brightly colored than the males. They lay several clutches of eggs in a season, moving quickly from one male to the next after laying each clutch. The males incubate the eggs and care for the young hatchlings. Female phalaropes are reminiscent of some of the male birds we discussed above, performing elaborate displays and showing off their beauty when ready to mate in order to convince any male to incubate each clutch. These examples are few and far between, but they support a general rule about mate choice—individuals who invest heavily in reproduction tend to be choosy about their mates, presumably maximizing quality of offspring over quantity, and those who invest little tend to be showy, competitive, and indiscriminate about their mates, aiming for quantity.

What about species in which both parents invest heavily in producing and caring for offspring? These examples will be most relevant to us, for all human cultures exhibit some degree of biparental care. Returning to my backyard house finches, females are not the sole childcare providers. Although females construct the nests, incubate the eggs, and feed the hatchlings, males pitch in in a variety of ways. They feed their mates during breeding time and egg incubation, and they help feed hatchlings and fledglings. This partnership may explain why males aren't bigger show-offs (they are quite dull compared to, say, the birds of paradise I mentioned earlier). Instead of showing off, male house finches spend their energy protecting and providing care for mates and highly dependent offspring. In addition, this partnership may explain why male house finches are also choosy. Studies have shown that males prefer older females to younger ones and those with some pops of color (although females tend to lack coloration, some females have a subdued version of the same coloration found on males). Thus, mate choice in house finches is certainly more complex than I presented earlier. Both sexes have preferences, and they aren't necessarily the same ones.

To make their story even more interesting, it has also become clear that house finches and many other bird species that were thought to be strictly monogamous are not. Both female and male house finches are known to mate with individuals other than their partner during a breeding season. Male house finches also guard their mates to prevent them from copulating with other males. So sexual conflict has also likely played a role in the evolution of reproductive strategies in this species.

Sexual selection operating on our human ancestors was undoubtedly also complex. Like house finch nestlings, human children require a huge amount of food and care. Raising a child to independence in our species requires more time, more skill, and more energy than it does in other primate species. Human mothers can't do it alone. Many primate mothers are the sole caregivers, whereas in our species, women require help. Other members of our communities contribute, whether it's a grandparent, aunt, older sibling, day care center, or school. And most relevant to our discussion of mate choice, human fathers invest heavily in their children (there is variation, of course, in what and how much they invest across and within cultures). Some degree of paternal care is observed in all human societies today, which suggests that it likely existed in our human ancestors in the past. Thus, we would expect men, like the male house finch, to have evolved to be somewhat choosy about their mates. Whether it's a for a one-night stand or a longer commitment, men today obviously have preferences and act on them.

Examples abound in pop culture and in our own lives. I have not-so-fond memories from my twenties of male friends using a "face-body" system to rate women they were interested in, with the obvious goal of hooking up with women with higher scores (I'll note that they did vary in how they weighted the face and body scores, a point I'll come back to later). Contrast these behaviors to those of male chimpanzees or bonobos, who will approach any female in heat who crosses their path, and it is clear that men in our species have a discerning eye. Women are choosy too though. Our human ancestors likely received help raising their children as I mentioned above, but they still shouldered much of the work, in particular during pregnancy and breastfeeding. Again, examples of female preferences today abound, but a recent one is the statistic that on Tinder, 80 percent of women swipe right on (select) only 20 percent of the men. Like the African widowbirds with artificially lengthened tails, a fraction of men on Tinder are getting most of the action because of shared female preferences for good-looking men.

Although I've focused on physical traits, preferences in women and men (and in other animals too) certainly go deeper than looks. Human preferences are also affected by cultural attitudes about beauty and sex, issues I'll tackle in the next section. In addition, sexual selection in our primate and human ancestors was certainly multifaceted, likely an interplay between mate choice (by both sexes), competition (within both sexes), and sexual conflict. Suffice it to say here, both women and men in our species are choosy about mates, consistent with the evolved role of both women and men in human family life.

The what

I mentioned male satin bowerbirds earlier, who win over choosy females by constructing impressive bachelor pads and decorating them with objects from their environment. But what exactly are females looking for when they shop around?

Our task would be straightforward if the bower functioned as a territory or nest, and females were simply looking for the best bower in which to incubate and raise their offspring. In many species, females do make their choice based on gifts, nests, protection, or other things that will directly impact their reproductive success. The scorpion fly illustrates this point: The female chooses a male based on the quality of the food items he offers her during courtship. Studies on humans across the globe have shown that resources and wealth do influence mate choices. All that said, bowers are not homes or nests—females build those at a different location. In many cases, including in satin bowerbirds, males provide nothing tangible to the female—no food, no protection, no nest, no care for offspring. And yet the female is still very choosy, basing her choice instead on the attractiveness of potential mates.

In a recent book about mate choice, *A Taste for the Beautiful*, UT Austin animal behaviorist and mate choice expert Michael

Ryan splits animal beauty into three types of traits: visual, auditory, and olfactory. Thus far, we've mainly focused on visual signals that appeal to the eyes. More precisely, as Ryan explains, attractive traits displayed by the courter are only *taken in* by the appropriate sensory organ in the individual being courted—for visual traits it's the eyes, for auditory traits the ears, and for olfactory traits the nose or other smelling organ—but then the information must be transmitted to the brain, which interprets it and decides what to do. So those visual traits we've been talking about are in fact appealing to the female *brain*. Back to our satin bowerbirds, the males are similar to other avian hotshots in using visual traits, such as their striking violet-blue eyes and shiny deep-blue plumage, to appeal to the brains of plainer-looking females. They also perform an intense courtship dance, flicking their wings and tail feathers up and down and hopping around. In the same way that dance movies are visually appealing to me, some females find these hopping wing flicks pretty enticing.

But more remarkably, male satin bowerbirds are architects and interior decorators as well. They build and decorate bowers to visually impress females. The bower consists of an upright entrance, resembling a hallway, that leads to an open flattened area on the forest floor. The hallway is constructed with twigs and branches, and the floor is decorated with items that are blue, an attractive color to females in this species. They use natural objects such as flower petals and berries to decorate but also man-made ones such as bottle caps and straws. They are fastidious about object placement, moving an item repeatedly until they're satisfied with the feng shui.

During courtship, a female enters the hallway and watches the male. She's probably visited the bower before—the first step in her decision-making process is visiting the empty bowers of multiple males. She revisits the ones she likes, and the second time around, she scrutinizes the male himself. While the female is in the hallway, the lustrous dark-blue male performs on his blue stage with his wing flicks and hopping. He'll take dance breaks by picking up his best blue items with his beak and showing them off to the female. If she's still undecided, she leaves and continues visiting other bowers. When she's finally chosen a mate, she returns to the chosen male's bower hallway and crouches down low, which signals to him that she's ready to mate.

All the signals I just described are visual ones. Mate choice in many species, especially in birds, is mediated by visual signals. Visual signals are less common in mammals, but mandrill monkeys, with their red and blue noses, red penises, and multicolored rumps, are quite visually striking, an observation that Darwin himself made in his discussion of sexual selection. Courters often combine different visual signals to impress, like the satin bowerbird's plumage, dancing, and bowers. And often a courting animal will send signals that appeal to multiple sensory organs at once. An important component of the male satin bowerbird's performance, in addition to dancing on stage, is his "singing," where he emits squawking, buzzing, and trilling noises to round out his performance. Back to my backyard house finches—males perform their visually appealing moves in the air while singing melodies, the more complex and longer the better.

Visual and auditory signals are often used together for courtship, but not always. If the courters aren't near those they're trying to woo, or they're wooing in the dark, they might use sound alone. Ryan studies Panamanian túngara frogs, which use a combination of two sounds in their communication with other frogs. They start with a long, high-pitched "whine" and add deeper, shorter "chucks" or honks at the end of the whine. A chorus of whining and chucking males reminds me of shooting sound effects in video games from the '80s. These frogs are an excellent demonstration of our rules on mate choice: It takes a female six weeks to get a batch of eggs ready for fertilization, whereas a male can fertilize one batch and relatively quickly move on to another. During the evening, when túngara frogs mate, males congregate in pools of water and try with their vocalizations to woo the few females nearby who are ready to mate. In clever laboratory experiments using loudspeakers to broadcast combinations of whines and chucks of different qualities, Ryan and his team discovered that females prefer the combination of whines and chucks together over either sound alone, and they prefer more and deeper chucks.

Like frogs, many mammalian species are active at dusk or at night, so mammals also use their voices to attract mates. Like the túngara frogs, female red deer prefer males with deeper roars. Like house finches, white-lined bats with more complex "songs" attract more females to their territories than those with simpler ones.

Also important in mammals are olfactory signals, the third type of beauty outlined by Ryan. Olfactory signals are an effective way to communicate in the dark and over long time periods. Any dog owner knows how important odors are to dogs, who obsessively leave their scents behind and obsessively smell the scents of others (including those directly in the behind). These odors convey information about potential mates and potential rivals, such as social rank and readiness to mate. Stallions sense when a female is ovulating by sniffing her urine, a process accompanied by a one-sided grimace that has been compared to the sexy sneers of rock stars like Elvis Presley or Billy Idol. Olfactory signals have attracted less attention than the stunning visual and vocal traits of birds, but a few experts have suggested that sexual odors in mammals are equally elaborate and important during mate choice.

Some of the clearest evidence for the role of odor in mate choice comes from studies of the MHC genes. The major histocompatibility complex (MHC) is a group of genes that play a role in vertebrate immunity. The genes in this complex direct the production of proteins that identify potentially dangerous invaders in the body, like bacteria and viruses, and alert our immune cells to come and destroy the invader. The MHC genes are the most variable genes in vertebrates, and for good reason: Having a greater variety of MHC genes means your immune system will more easily recognize and fight off more infections. While these genes play a critical role in immune function, what do they have to do with odors and mate choice? Studies in stickleback fish and rodents have shown that individuals that have more similar MHC genes smell more similar to each other and vice versa. The odors emanate from bodily secretions like sweat and urine, and animals use them differently depending on the context. In some bird and rodent species that communally raise young offspring, individuals prefer to commune with other family members (to remind yourself of why, you can go back to our discussion of kin selection in the introduction). Studies have shown that kin recognition in these species is mediated by MHC-based odors—individuals seek out those who smell the same.

On the other hand, when choosing a mate, individuals in many species, including mice, salmon, stickleback fish, and humans, seek out those who smell different, presumably to make healthier offspring with a greater diversity of MHC gene versions and/or to avoid inbreeding. Some illuminating studies have been done in humans. In the "sweaty T-shirt" experiment, the MHC genes of several college students were sequenced to determine what versions they had. The men were asked to wear the same T-shirt for a couple of days, and the women were then asked to rate the attractiveness of the men based on the smell of the T-shirts. Women preferred the smell of T-shirts worn by men with different MHC genes than themselves. Interestingly, if a woman was taking birth control, the opposite effect was observed—women preferred the smell of MHC-similar men. The researchers hypothesized that this discrepancy is due to the fact that birth control hormones stop us from ovulating by mimicking the hormonal milieu during pregnancy, and pregnant women, like communal birds and rodents, prefer being around family for help.

Another line of evidence supporting the role of MHC-based odors in human mate choice comes from a well-studied community of Hutterite people in South Dakota. Hutterites migrated from Germany to rural South Dakota in the late nineteenth century, and researchers have been studying them for decades—we have excellent genealogical and genetic information on this small community. Studies have shown that Hutterite couples who have similar MHC genes have recurrent miscarriages, indicating that the costs of mating with someone with a similar MHC profile are experienced soon after conception. In fact, this observation has been made in populations across the globe, and it may have evolved as a screening mechanism in the mother to discard poor-quality embryos. Other studies have shown that Hutterite individuals with similar MHC genes get married much less frequently than you would expect by chance. The results from the sweaty T-shirt and other experiments just mentioned would suggest that Hutterites are preferentially choosing mates with complementary MHC genes based on their smell.

Most readers will know from experience that all three

types of signals I've been talking about—sight, sound, and smell—play a role in human attraction (and repulsion—I briefly dated someone in my twenties who must have had an identical MHC profile to myself). Entire industries are built on this presumption—plastic surgery, high fashion, and cosmetics enhance our visual appeal; perfumes and colognes enhance our olfactory appeal. It is also probably obvious to readers that we don't all share the same preferences. While I love visually appealing dance movies, I have extremely sensitive ears—I'm one of those people who plug their ears when a fire engine passes by. So I have low tolerance for musicals and for very chatty men—I've always preferred the quiet type. I mentioned those friends from my twenties with the "face-body" rating system who varied in their preferences—some weighted the face score over body, and others body over face.

Personal anecdotes aside, while early studies of animal mate choice assumed that female preferences are immutable, more recent work has shown that individuals in many species also vary in their preferences, in particular those who evaluate multiple signals at once. Female cowbirds evaluate both visual and auditory traits during mate choice, but their specific preferences vary based on how sensitive the visual and auditory processing parts of their brains are. Female satin bowerbirds vary in their preferences by age: Young females, who are more threatened by aggressive male display, make their decisions based on appearance, whereas older females choose based on the intensity of the male's song and dance. A fascinating observation of males in these same species is that the most successful ones vary their display according to how the female is responding to courtship. If a young female satin bowerbird cowers in response to an aggressive visual display, some males will dial it down a few notches. When courting females, these males seem to have high emotional intelligence (also known as the emotional quotient, or EQ).

A growing body of work across many vertebrate species supports what we know to be true in humans: Individuals vary in their preferences. This variation likely contributes to maintaining variation in the display traits themselves, a requirement for sexual selection to continue operating. One nagging issue from our earlier discussion of human breasts is why there is variation in breast size at all. One explanation is that men in the past varied in their preferences for smaller or larger breasts or in their interest in breasts versus other traits, which also seems to be true of men today.

In his manifesto on mate choice, *The Evolution of Beauty*, Yale ornithologist Rick Prum argues that many of the traits that develop at puberty in our species, such as permanent breasts, a large, dangling penis, and pubic and underarm hair, evolved by sexual selection, in some cases by male mate choice, in others by female mate choice, and still others by mutual mate choice, in which both sexes share the same preferences. More controversially, he supports Darwin's hypothesis that many of the physical differences between humans across the globe were driven by differences in cultural mating preferences.

While some physical differences have clearly arisen by natural selection, such as dark skin at equatorial latitudes (for protection against skin cancer) and light skin at higher latitudes (to enable the production of vitamin D), most traits cannot be easily explained as adaptations to specific environments. This hasn't stopped the adaptive explanations from pouring forth to explain everything from differences in hair texture to differences in penis and breast size. But an alternative explanation, first proposed by Darwin and more recently advocated by Prum, is that cultural differences in mating preferences had a top-down effect, whereby they drove the genetic changes underlying most of the physical differences between human groups.

These ideas haven't been formally tested yet, but Prum discusses how this might have worked by looking at the Khoisan group in southern Africa, which prizes large accumulations of fat in the buttocks area. According to Prum, the distinct pattern of fat deposition is unlikely to be an adaptation for the specific environment of the Khoisan people. More likely, it began with a slight preference for bigger behinds, which kicked off a feedback loop between the preference and the trait, leading to a strong preference and an exaggerated trait. More generally, Prum makes an ardent case for the role of mate choice in human evolution, influencing not only superficial, physical attributes but also social and behavioral traits that paved the way for the unique evolutionary trajectory and success of our species.

Indeed, while sight, sound, and smell clearly play a role in human attraction, they are certainly not the whole story. When we think about the people we love, personality traits are often the first thing to come to mind. Perhaps the deep voice, high cheekbones, or "10" body attracts us initially, but traits such as kindness, humor, curiosity, and high EQ are also important in whom we choose for the longer term. Psychological studies on the interplay between physical and personality traits in sexual attraction have shown that while people often agree on their original opinions of attractiveness (reflected in the statistic I mentioned above on fewer men on Tinder getting all the right swipes), our perceptions of attractiveness change as we get to know others. In other words, personality heavily influences to whom we are attracted. Like physical traits, Prum argues, personality traits in humans evolved by mate choice selection. Our human ancestors preferred individuals who were a little kinder, funnier, and more empathetic, and over time, this led to the persistence and elaboration of these personality traits in humans. Just like long tails in widowbirds and the preferences for them, a feedback loop between personality preferences and the personalities themselves resulted in many of the traits that we find attractive in prospective mates today.

Again, these ideas have not been formally tested, but we may see hints of this process operating in some of the animals I've been talking about. I mentioned above that female satin bowerbirds prefer males that can read social cues, modifying their courting behavior based on how the female is responding to their display. This high EQ bird trait likely evolved by mate choice—the males who better read the cues attract more of the females. Back to my backyard house finches, we've discussed how bright red males are preferred by females, but one study showed that the duller red males can still compete successfully for mates. Males who are more social, flocking with multiple groups during the winter instead of just one, can compensate for their drab appearance and are just as successful as the brilliant red ones. Duller males are four times more likely to hang out with multiple groups than the bright red males are, presumably to enhance their appeal. So just as some humans are attracted to "social butterflies," it appears that some house finches are as well.

The when (and where)

Even the amateur comedian knows how important timing is when telling a joke. The knock-knock joke featuring the interrupting cow has been a staple in my house for years. My kids now use characters other than cows in the joke to keep their material fresh. Their "mom" version reveals the type of daily conversations I have with my kids:

"Knock knock."

"Who's there?"

"Interrupting Mom."

"Interrupting Mom wh—"

"WHO PEED ALL OVER THE BATHROOM FLOOR?!"

Just as timing can make or break a joke—my five-year-old still has difficulty with the timing of that joke, waiting too long to deliver the last line—timing is also critical during mating decisions.

Back to our prototypical choosy female, if she's an animal with a backbone, she has a reproductive cycle roughly resembling ours. Under control of hormones from the brain, the first part of the cycle is characterized by estrogen production and egg development in the ovaries. Once eggs are ready, one or more are released. In mammals, ovulation is followed by progesterone production by the ovary, which primes the uterus for pregnancy. As discussed in the last couple of chapters, while the basic components of the brain-ovary reproductive axis are similar across vertebrates, one major variable is what triggers the start of the cycle. In many animals, environmental cues (like the length of the day) tell the brain to send the signals to initiate egg development; in others, like ourselves, the cycle is a spontaneous loop that feeds back on itself, requiring no external cues to keep it going. In the majority of vertebrates that make choices about mates, mating decisions only happen when the reproductive system is ready for fertilization (i.e., around the time of ovulation). For all those vertebrates whose cycles are tied to the seasons, mate choices are made at the time(s) of year when ovulation occurs.

Studies on white-crowned sparrows nicely illustrate this point. These birds only breed in the spring, so a female's estrogen levels become elevated during the spring months as her ovaries get eggs ready for ovulation. As in many birds, females in this species choose males based on their song. We discussed in the last chapter that some behaviors, like sex, activate a reward system in the brain, motivating individuals to engage in these behaviors. This activation also happens during mate decisions. Darwin was the first to hypothesize that the sensory pleasure experienced during sex is similar to that experienced when evaluating beautiful traits in potential mates, and we now have evidence to back up his idea. When a female white-crowned sparrow hears the melodious song of a male, reward areas of the brain are activated, encouraging the female to go for it, but this only happens during the spring. If a female hears a male sing in the winter, which he does to defend his territory against other males, she attacks him. This difference in behavior in the winter versus spring has been linked to estrogen levels. Only when a female's estrogen is high does the reward system become activated in response to the song. If her levels are low, watch out!

Humans are unusual in the animal world in that sexual activity is not limited to the short window of time during which fertilization is possible. Sex happens more frequently in humans, decoupled from its primarily reproductive function in other animals. Prum's explanation is that sexual pleasure and the organs involved in generating it were also elaborated on in humans by the process of mate choice selection. He argues that the capacity for multiple orgasms in women is one such elaboration. Nevertheless, there is plenty of evidence that sexual interest and activity change over the course of the human menstrual cycle. As in white-crowned sparrows, women are more interested in sex (and initiate sex more) in the days leading up to ovulation. They are also more restless, and their senses are heightened, including their sense of smell— perhaps to better detect the MHC-based odors discussed earlier. Remember that in humans, mate choice is complex, with both women and men making choices. Consistent with this, another behavior observed in women at different points in their menstrual cycle is that they pay more attention to their appearance at midcycle, wearing more fashionable clothing, revealing more skin, and wearing more makeup.

One interesting effect of time on mate choice is what happens when time is running out. Back to the Panamanian túngara frogs, it takes a female six weeks to get her eggs ready, and she has one night to get them fertilized. If she doesn't, the eggs ooze out of her body and are wasted. At the beginning of her one night of love, she prefers the whine and deep chuck mating call combination that we discussed earlier. But as the night wears on, if she has not yet mated, her standards change as she approaches the time that her eggs will ooze out. She is more likely to respond to "unattractive" synthetic mating calls than just a few hours earlier. This phenomenon of changing standards of beauty with time was observed in a classic study from the late '70s on women and men at bars (the study was repeated more recently to control for the effect of alcohol, and the finding was still supported). At the beginning of the night, people rated the attractiveness of potential mates in the bar. As it got closer to closing time, the same people were given much higher ratings. In other words, people's perceptions of beauty changed as time was running out to find someone to go home with. This "closing time" effect also operates on a longer timescale, coming into play toward the end of one's reproductive life. As individuals near the time when they won't be physiologically able to reproduce anymore, they often become less choosy about their mates. Changing standards with age have been observed in a range of animals, including roaches, guppies, crickets, and fruit flies, and it has been suggested in humans. A study of over eight hundred women of different ages found that women in their thirties and forties have more sexual fantasies, are more willing to have intercourse, and have more intercourse than younger women, suggesting that women become less choosy as they approach menopause.

While mating decisions in our species are sometimes made in bars, where were they made in our human ancestors as sexual selection and mate choice were shaping human biology? We don't know exactly how our ancestors congregated in their environments, but we do know that the way individuals in a species get together across space heavily influences the process of sexual selection. This clumping of individuals is itself influenced by ecological and social factors such as where the food is and how family members associate with each other. Female mandrill monkeys form social groups consisting of related females and their offspring. Because female mandrills clump together in space (and they reproduce seasonally so are clumped together in time too), male mandrills have evolved a strategy of sexually monopolizing a group of females to the exclusion of other males. Sexual selection by male-male competition has dominated in these animals. But when individuals are more physically dispersed in the environment, which happens for a number of reasons, a reproductive strategy often taken is monogamy, in which males and females mate more or less exclusively over an extended period of time, and they often both care for offspring. We will revisit these topics in future chapters. But now I want to turn our attention to the lingering question you may still have as you've been reading about mate choice: Why are we choosy about our mates in the first place?

The why

Let's return to my backyard house finches, who prefer brighter red plumage, more elaborate songs, and more sociable males. As we've been discussing, it is now widely accepted that house finches, and individuals in many vertebrate species, have mating preferences and act on them. It is also agreed that the action of mate choice is a powerful evolutionary force, driving both the evolution of desired traits and the desires themselves. The disagreement comes when trying to explain why the desires exist to begin with. Why do female house finches prefer red, sociable males who make beautiful music? What's in it for them?

The answer to this question is obvious when individuals choose mates based on a direct benefit, something immediate and tangible like food, a nest, or protection. The scorpion fly females I mentioned earlier prefer the males who bring them the best food. But in many cases, all males bring to the table are their sperm. Even in those cases, though, females can be extremely choosy.

Darwin's answer to this question was that animals have subjective aesthetic preferences, a "taste for the beautiful," as was mentioned earlier. Darwin invoked this humanlike sense of aesthetics to explain the evolution of animal beauty. Females prefer mates displaying the most beautiful ornaments simply because the ornaments are pleasing to them. Once females in a species have preferences and start acting on them, they beget daughters who share their preferences and sons who share the preferred trait with their dad. If females in a species generally agree on what they like, then the preferences and the traits take over in a population and can become more exaggerated over time. In addition to the sensory pleasure experienced immediately while evaluating and choosing an attractive mate, a female indirectly benefits from her choice because her sons will be considered attractive to females in the next generation (this is sometimes referred to as the "sexy sons" hypothesis for mate choice). So she will have more grandchildren, great-grandchildren, and so on because her male descendants are more attractive. Darwin did not know modern neurobiology, such as the existence of specific pleasure and reward systems in the vertebrate brain (like those activated when white-crowned sparrows evaluate courtship songs), but the presence of these ancient brain systems is consistent with his idea that animals have the capacity to experience sensory pleasure and make social decisions based on it.

Darwin's explanation of animal aesthetics is spot-on to some biologists, including Prum, who is on a mission to revive Darwin's original version of sexual selection by mate choice. But the notion that animals have a humanlike sense of aesthetics that has the power to drive their own species' evolution, as opposed to more accepted external factors like climate, predation, and food availability, is considered unlikely by many, including Darwin's contemporaries and many of his successors. An alternative explanation is that preferences exist for a reason—specifically, the preferred traits are indicators of the genetic quality of potential mates (the "good genes" hypothesis for mate choice, which was introduced earlier). A male in better condition because of his better genes—such as his immune genes—will produce more beautiful ornaments. A female indirectly benefits from her choice of the most beautifully ornamented male because her children (both females and males) will be genetically superior to the children she'd have with a less beautiful male, and they'll go on to produce more robust children themselves. So beautiful traits evolve so that females have a reliable way to choose the superior males. This perhaps makes intuitive sense and is more consistent with Darwin's adaptation by natural selection—only the "fittest" survive and reproduce.

But the evidence supporting the good genes hypothesis is thin. One of the textbook examples is red and orange body coloration in a range of species, including house finches. Several studies have attempted to establish a link between brighter red color and higher-quality individuals, such as that in stickleback fish: Males with bright red bellies are more attractive to females and they seem to resist parasites better. However, after decades of research across a range of species, it is unclear how frequent the relationship between red color and good genes is, and the mechanisms linking the two traits are still much debated. Moreover, red coloration is a relatively simple trait, compared to, let's say, the multicolored and exquisitely patterned plumage of the peacock. The good genes hypothesis requires that each dimension of a display evolved because it was a better indicator of good genes, an assertion that Prum thinks impossible for some of the highly complex avian displays in nature. All that said, I've already discussed a well-supported example of a trait that allows choosers to find preferred genes: MHC-based odors. Individuals in many species are known to choose mates based on these odors, which don't reflect superior genes exactly but rather complementary ones to the chooser. In some cases, then, the right genes are what we're after, and we use smell to find them.

Yet another idea on why females have specific mating preferences is that their brains are already wired to prefer certain signals. The males evolve traits to exploit these preexisting biases or preferences. We've already discussed an example—túngara frogs and their preference for deep chuck mating calls. Initially, it was thought that deep chucks were preferred because they were correlated with direct benefits or good genes. Males who produce deep chucks are larger, larger males fertilize more eggs, and the offspring of larger males appear to have better survival. But a careful study of close relatives of túngara frogs revealed an alternative explanation. Even though the majority of these frog species do not produce any chucks in their calls, one of their two hearing organs is perfectly tuned to hear deep chucks. As Ryan's team argued, túngara males that fortuitously evolved the ability to chuck were more successful than the chuck-less males, exploiting this preexisting bias for deep sounds.

A more intuitive example is the preference in guppies for the color orange. Guppies often feed on orange fruit that falls into the rivers in which they live; females in many guppy populations also prefer males with brighter and larger orange spots. A series of elegant experiments showed that the preference for orange fruit likely arose first, followed by the exploitation of this preference by males to attract females. In other words, the guppy brain first evolved an attraction to the color orange to facilitate more efficient foraging; then, males exploited this preference to attract females during courtship. While this possibility hasn't been studied in house finches to my knowledge, it is interesting to note that house finches feed on a variety of seeds and berries, their favorites being the reddish mulberry and cherry. In the most general sense, these preexisting biases are likely why birds use bright colors to attract mates, whereas mammals, many of which are not active during the day and thus don't have the same visual acuity as birds, use auditory and olfactory signals instead.

Last but not least, it is possible that some combination of these mechanisms—direct benefits, sexy sons, good genes, and preexisting preferences—better describes the evolution of some mating preferences and their corresponding traits. One of the original mathematicians who described the sexy sons process, Ronald Fisher, developed a two-part model that starts with a preference for a trait that indicates good genes. But once the process starts, it takes on a life of its own (the sexy sons process is also referred to as Fisherian "runaway" selection). When the preference and preferred trait become inherited in the same individuals, a feedback loop ensues, in which the preference and trait become more and more exaggerated in future generations. The trait no longer provides any honest information about the genetic quality of the courter— the desired trait becomes unhinged from the good genes. According to Fisher (and other statisticians who worked on this problem), a mathematical inevitability is that the desire for the sexy trait, the taste for the beautiful, takes over.

———

So what does all this mean for us humans? Back to the title of this chapter, "What's Love Got to Do with It?" Romantic love has in fact been studied by neurobiologists, anthropologists, and psychologists who have described it as an evolutionary elaboration of the brain system for mate choice. The pleasure/reward systems activated during mating decisions in birds and mammals are also activated during the experience of romantic love in humans. The intense attraction we feel toward a person with whom we're falling in love may be a prolonged and intensified version of what a house finch experiences when a male from her social group courts her with song and dance.

Just as romantic love has the power to shape and transform our individual lives, mate choice in our evolutionary past was a powerful force, molding human anatomy and behavior in a myriad of ways. While the ultimate reasons behind our preferences—direct benefits, sexy sons, good genes, preexisting biases—will continue to be debated for decades to come, there is no question that mate choices in the past transformed our species physically and socially. Some universal human traits, such as permanent breasts and a prominent penis, likely evolved by mate choice selection. Many of the physical differences between geographic groups, and social/personality traits as well, may have also evolved by sexual selection, although further investigation is needed. Our preferences and choices today may provide clues as to how sexual selection is currently operating in human populations—a topic beyond the scope of this chapter—but they also give us hints about our evolutionary past. This is a favorite topic in the field of evolutionary psychology, which attempts to explain all human behaviors, including mating preferences, as past adaptations. The field tells us that women across the world prefer men with greater resources and that men, who are sexually profligate, care more about looks (such as breast size) because they advertise fertility. These explanations are contentious, and even if they gained unanimous support, they can only be a sliver of the full story. As we've learned from animal studies, mating preferences are not fixed but vary across individuals and also in the same individual across time. Moreover, individuals often make their decisions based on multiple traits, and they don't place the same value on each trait. Clearly human preferences are diverse and multifaceted, and thankfully so. Otherwise we'd all fall in love with the same people.

As I write about mate choice, I can't help but reminisce about the time my husband and I chose each other for the long term. We had a whirlwind long-distance romance while he was living in NYC and I in San Francisco. Among the many reasons I chose him: He made me laugh and didn't take himself too seriously; he was tall, dark, and handsome; he was well read and loved to travel; I thought he would be a supportive partner and father. His proposal on the Hudson River could have been a script for a Hollywood romance—until halfway through his speech. Everything was perfect until that point: The setting was romantic, and he was so earnest while telling me what he loved about me...my smile, my looks, my work ethic, my curiosity. But then he said something along the lines of "I'm choosing you for your good genes." As unromantic as that was to me at the time (it kind of killed the mood!), I can appreciate now that his honest speech reveals the many factors that go into these choices, and it hints at the diversity of evolutionary mechanisms that are involved.

CHAPTER 6

The Fraught Path of Pregnancy

Of all human connections, that between a mother and her baby is the closest. I'm not referring to the breastfeeding newborn nestled in the crook of your arm or the toddler who won't sit anywhere except ensconced in your lap. I'm referring to the baby who hasn't been born yet. A pregnant mother shares every morsel of food with her developing fetus, every breath of air. When a woman experiences high levels of stress during pregnancy, she shares her stress hormones with her baby. A woman who uses alcohol or drugs during pregnancy shares those substances with her fetus too. For better or for worse, a mother and fetus have the most intimate of human connections during those forty or so weeks of pregnancy.

In an evolutionary sense, this connection serves both mother and fetus. A fetus's protection, nourishment, and health through pregnancy benefit both parties: The mother passes on her genes through her child, and the fetus will eventually pass on its genes through itself. Evolution should have polished this relationship to perfection. So one of the puzzling things about pregnancy in our species is just how many things can go wrong. Some are lucky to have easy pregnancies that end with the birth of a healthy baby, but the road for many can be extremely (and sometimes heartbreakingly) difficult. That road might include infertility, recurrent miscarriage, ectopic pregnancy (a pregnancy that takes place outside the uterus), nausea, hyperemesis gravidarum (excessive nausea and vomiting), anemia, gestational diabetes, high blood pressure, preeclampsia/eclampsia (extremely high blood pressure), preterm labor, postpartum infection, or postpartum hemorrhage. From the evolutionary perspective, successful reproduction is the point of it all, so you'd think that natural selection would have worked out more of the kinks.

The principal question we'll address in this chapter is why these vulnerabilities exist. Why did pregnancy evolve in such a suboptimal way, exposing us to so many complications?

First, we'll need to get the evolutionary lay of the land. We'll discuss how various mammals grow their babies—from egg-laying monotremes to pouch-forming marsupials to the most successful group, placental mammals, which rely on a complex placenta for the transfer of nutrients and oxygen directly from mother to developing baby during

pregnancy. We'll compare the key players of pregnancy—
the fetal placenta and the maternal uterus—across
mammals to understand the major events involved in the
evolution of pregnancy. These comparisons will reveal that
pregnancy evolved under conditions of cooperation *and*
conflict between mother and fetus. The notion that mother
and fetus have to cooperate during pregnancy is intuitive,
but the existence of conflict is not. Remember from previous chapters that the potential for conflict exists because
mother and child are not genetically identical. Genes in the
mother and fetus agree that the fetus should survive and
reproduce, but their interests diverge if the fetus gets too
greedy, compromising the mother's ability to have more
offspring. Maternal genes may respond to fetal greediness
by becoming stingier, balancing a mother's own needs with
those of current and future children. These conflicting
interests explain the propensity in humans for complicated
pregnancies.

I'll also reveal a twist to the story of pregnancy in
mammals, which involves conflict between completely
different actors. The evolutionary negotiations between
mother and fetus were themselves enabled by parasitic
DNA elements that have been in conflict with our genomes
for hundreds of millions of years. We'll discuss what these
parasitic elements are and how mother and fetus have
harnessed them in their own conflict and cooperation in
the womb.

A short history of baby making in animals

Many women who have been through pregnancy will agree that it is like no other experience. Growing a human being inside my body was at times bizarre, amazing, trying, and deeply satisfying. Having been through a full pregnancy four times, I can also say that it is idiosyncratic, no two experiences being quite alike. I spotted through most of my second pregnancy, was nauseous through most of my fourth, and had three that went long and one that went short.

During my investigations into the evolution of pregnancy as a biologist, I have had some of the same reflections on baby making across the animal kingdom that I had through my own pregnancies—bizarre, amazing, idiosyncratic. To provide the needed context for understanding human pregnancy, let's look at how different animals make and support their developing babies.

For vertebrates (animals with a backbone) to move from water onto land, they had to evolve limbs, the topic of *Your Inner Fish* by paleontologist Neil Shubin (one of my favorite nonfiction science books). But they also had to change how they made their babies. The first land-dwelling vertebrates were similar to today's amphibians (frogs and salamanders), able to move on land but needing to reproduce near water. Most female amphibians (and fish) release eggs with a jellylike coating that get fertilized by sperm outside their bodies. These eggs have yolks to nourish developing babies, but they don't have much protection from harsh conditions, predation, or drying out (which is why fertilization must happen in water). Females in these species must release hundreds or thousands of eggs at each ovulation since the vast majority don't make it—either they don't get fertilized, they don't complete development, or they don't reach reproductive maturity.

The earliest amniotes took advantage of two major innovations: the amniotic egg and internal fertilization. The amniotic egg has four specialized membranes—the amnion (surrounding the embryo), yolk sac (storing the yolk), allantois (storing waste), and chorion (enclosing all of the above)—and an eggshell. These egg features freed amniotes from reproducing in the water, providing a solution to drying out and to some of the other problems faced by amphibians, because amniotic eggs can be buried or hidden. The evolution of the eggshell required internal fertilization, though, since sperm can't penetrate an egg with a shell. As I mentioned in chapter 4, the first amniotes evolved a penis that allowed sperm to fertilize eggs within the female's reproductive tract. The female would have held off wrapping her eggs in a shell until after fertilization, as we observe in egg-laying amniotes today, including many lizards. After egg laying, the developing amniote fetus within its shell was relatively protected, fed off a large yolk, and could exchange gases and safely store waste until hatching.

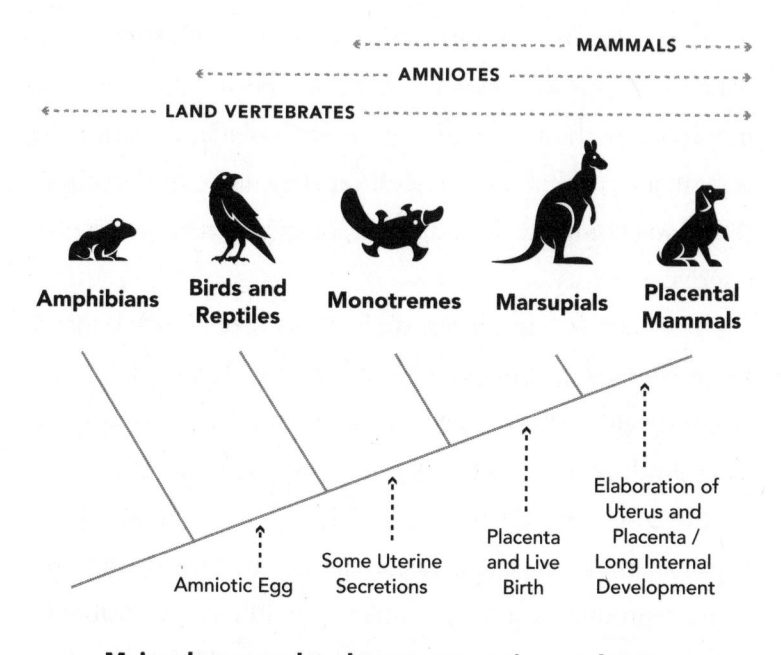

Major changes related to pregnancy in vertebrates.

Amniotic eggs radically transformed life on land, allowing amniote vertebrates to venture far from water to live in new habitats. But developing amniote babies were still vulnerable, even in their eggshells (amniote eggs are a favorite food of many predators, including human predators—think about that omelet or scrambled eggs you had for breakfast this morning). For some species, a safer option than laying fertilized eggs would have been developing offspring inside the body and giving birth to live babies. The transition from egg laying to live birth happened over one hundred times in vertebrates, and not just in amniotes. Live birth evolved independently in many fishes, amphibians, and reptiles (although to be clear, most species in these groups still lay eggs). In lizards and snakes, live birth is associated with living at high altitudes and low temperatures, suggesting that one environmental pressure for the transition to live birth is the cold. Pregnancy and live birth also evolved once in mammals. The independent, repeated evolution of live birth in vertebrates tells us that it must be a useful reproductive strategy. The only vertebrate group in which it has never evolved is birds.

So how does live birth evolve from egg laying? There is no one answer to this question, because it happened differently every time it occured. In some species (e.g., many live-bearing fishes), the developing baby just feeds off its yolk and hatches while still inside the mother. In others, the mother's reproductive tract has glands that actively secrete nutrients that are consumed by the developing baby; either they are taken in directly through the mouth, gut, or skin or are first absorbed by a placenta that then passes nutrients and oxygen on to the baby (more on the placenta in a minute). In the most macabre form of live birth, found in some species of sharks, babies obtain the necessary nutrients for development by eating their siblings in utero. These sharks take cannibalism to a more gruesome (and personal) level.

The placenta, a specialized structure that allows for the exchange of nutrients and gases between the mother and developing offspring, can be found in some fishes, amphibians, reptiles, and of course mammals, including ourselves. No two placentas look the same, and different groups came up with their own ingenious ways for babies to take in more nutrients from mothers during gestation. In some fish, a simple placenta evolved from the sac that surrounds the yolk; in others, it evolved from the sac that surrounds the embryonic heart. Live-bearing amniotes took advantage of the additional membranes in the amniotic egg to make their placentas, but for the placenta to work, they had to get rid of, or at least thin out, the eggshell. Many live-bearing reptiles use two of those membranes (the chorion and allantois) to make their placenta, and it starts working after the eggshell dissolves in utero. In these cases, the placenta nuzzles up close to the lining of the mother's reproductive tract, but it doesn't invade or breach maternal tissues. The placenta simply opposes the uterine lining and expands the surface by which the developing baby can take in nutrients released from the mother and exchange oxygen and carbon dioxide. The mammalian placenta, on the other hand, wins the prize for figuring out a way to tap directly into the mother's blood supply rather than waiting around to soak up whatever the mother is willing to release. Shark babies might be cannibals, but some mammalian babies are like blood-sucking parasites. How did this happen?

Remember from previous chapters that mammals can be divided into three groups: monotremes (the platypus and echidna), marsupials (kangaroos, opossums, etc.), and eutherian mammals, often referred to as placental mammals (most of the species you think of as mammals, including elephants, bats, dogs, mice, and humans).[1] If we think about mammalian relationships as a tree, the earliest mammals evolved from a reptile-like amniote over 200 million years ago, and they represent the trunk of our tree; the monotreme line branched off from the trunk about 180 million years ago; then the trunk split into marsupial and placental branches about 160 million years ago. No group is more evolved than the other, as they are all still living on the planet today. But placental mammals have been more successful as a group, with about 5,000 placental species still in existence today, compared to 5 monotremes and 250 marsupials. Evolving a complex placenta was clearly a winning strategy.

We think the earliest mammals on the trunk of our mammalian tree laid eggs, because the earliest branching mammals—the monotremes—still lay eggs and because most amniotes outside the mammal group (such as birds and most lizards and snakes) lay eggs. These eggs were likely covered in a leathery, permeable shell, like those of the platypus. The platypus mother lays two eggs at a time and incubates them for about ten days until hatching of the

---

[1]   Marsupials have a simple and short-lived placenta, but before this discovery, the third and largest group of mammals—placental mammals—were named based on this structure, which was thought to be unique to the group. After the marsupial placenta was described, placental mammals were renamed eutherian mammals. To keep things simple here, I'll refer to eutherian mammals as placental mammals, the mammals that make a complex and invasive placenta.

immature "puggles" (most adorable name for a baby animal ever). Then the mother suckles them for about four months until they become covered in fur and can swim. But before egg laying, the developing platypus spends several weeks in the reproductive tract of its mother, during which time it takes in nutrients secreted by the mother's uterine glands that diffuse across the eggshell. So although our earliest mammalian ancestors laid eggs like the platypus does, the first steps toward live birth may have been taken in these animals with the mother releasing nutrients and the fetus taking them in through the eggshell.

Marsupials have an unusual mode of reproduction. A marsupial embryo remains in the mother's reproductive tract, in its eggshell, during the earliest stages of development. As with the platypus, some secretions from the uterus are taken in through the eggshell, but the marsupial hatches out of the egg while still in the uterus. After it hatches, a simple placenta briefly attaches to the lining of the mother's uterus and facilitates the exchange of nutrients and gases. In the marsupials that have been investigated, the placenta is derived from a fusion of the yolk sac and chorion, and it is usually short-lived. Recent work out of my PhD lab showed that in gray short-tailed opossums, the total length of pregnancy is about fourteen days, but the placenta doesn't attach till day twelve. In other words, the placenta lasts for only two days before opossum "joeys" are born. Birth is thought to happen so quickly after placental attachment in this species because digestive enzymes that degrade the eggshell irritate the uterine lining, resulting in a maternal inflammatory reaction that causes labor. All marsupial joeys are extremely immature at birth. They look like tiny, helpless, hairless embryos with humongous arms and mouths—all the better to crawl to the teat and drink milk with, my dear! After birth, they need to spend a lengthy time suckling and living in a pouch or other protected area on the mother to complete development. The gray short-tailed opossum spends about eight weeks suckling after gestating for two weeks. A mouse, in comparison, which belongs to the third group of mammals—the placental mammals as I'll discuss next—gestates and suckles for the same amount of time, about three weeks. Milk in the marsupial, as in the monotreme, is the main source of nutrients needed for development.

Placental mammals have a placenta that is more complex, invasive, and longer lived than the marsupial placenta. Their placenta is so complex that it easily qualifies as an organ, requiring a lengthy period of development (the entire first trimester in humans) to be fully functional. In placental mammals, the eggshell never forms, and the fetus stays in the womb for an extended time, usually longer than the length of the mother's reproductive cycle. For small mammals, like mice, pregnancy lasts several weeks, but for larger mammals, like elephants and whales, it lasts almost two years. This is an amazing feat, given that the fetus is not genetically identical to the mother, which in other circumstances (like organ transplants) provokes the mother to reject the foreign object. These longer gestations allow developing placental mammals to complete more of their development while protected in the mother's womb. Placental mammals are born much more advanced than monotremes or marsupials, and in many cases, newborns can pick themselves up after birth and start going about their business. Zebras and giraffes can walk within minutes of being born.

Comparison of a newborn marsupial (kangaroo), monotreme (platypus), and placental mammal (dog).

In all placental mammals, the placenta is derived from the chorion and allantois—two of those membranes of the amniotic egg. The umbilical cord, which connects the fetus to the placenta, is formed from the stalk of the amniote allantois that emerges from the fetal gut. This is a great example of nature taking something that evolved for one reason—these membranes are important for waste storage (allantois) and gas exchange (allantois and chorion) in the amniotic egg—and tweaking it to make something new.

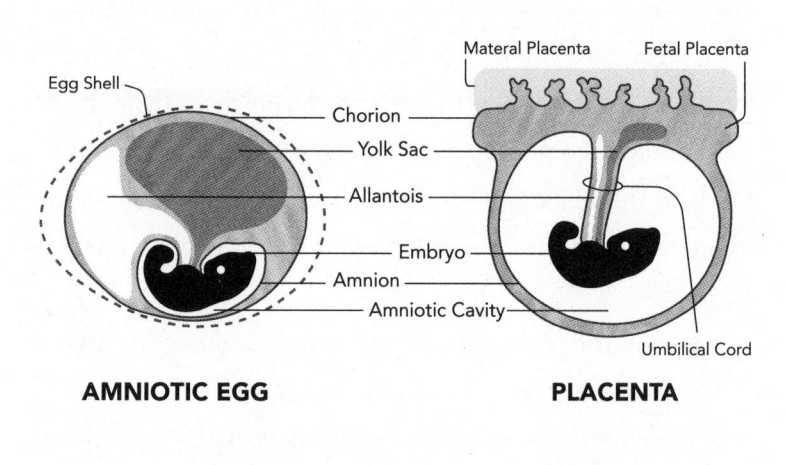

Comparison of membranes in the amniotic egg to those in the mammalian placenta.
Illustration based on figure from *Principles of Life*, 3rd edition by D. M. Hillis.

One of the key changes in mammals was a distinct cell type that emerged from the chorion—the trophoblast—that enabled the placenta to become invasive. In humans, trophoblast cells squeeze through the most superficial layer of cells of the uterus, and a subset of these cells can invade even deeper into the lining, hunting around for blood vessels. When these specialized trophoblast cells find blood vessels, they rip them open, widen them, and replace the vessel walls so the mother can't constrict them. The rest of the placenta then bathes in the blood that is released, providing the fetus with a steady supply of nutrients. This complex, invasive placenta did not evolve in a vacuum; the mother's uterine lining and her immune response had to evolve alongside placental tissues to deal with the invasion. On the maternal side of things, a new uterine cell type evolved too, the decidual cell. These cells have many functions, including controlling how deeply those trophoblast cells invade. The interaction between maternal decidual cells and fetal trophoblast cells is so intimate that some biologists consider the placenta to be an organ of both fetal and maternal tissues.

Uterine and placental tissues are so intimately connected during pregnancy that when you deliver a placenta after the birth of a child, those decidual cells get delivered too. Decidual cells are like leaves—they fall out of the uterus at birth (and during menstruation) the way leaves fall off a deciduous tree in the autumn. When I delivered the placenta after my first son was born, my obstetrician immediately asked if I wanted to see it, knowing that I was studying the organ for my PhD. It resembles a meaty, bloody, spongy, veiny disk, not quite the size of a baby but larger than you'd expect (almost a foot in diameter and weighing about a pound). The most evocative analogy I've come across to describe the human placenta is a "mop in a bucket of blood." The mop is the fetal placenta, which isn't a smooth disk but is covered in projections called villi that increase the surface area by which it can absorb nutrients. The bucket is the uterus, and the blood is maternal blood released by the uterus at the hands of those invasive trophoblast cells.

At the risk of sounding geeky, I think the placenta is the coolest organ around. First of all, a placenta evolved independently in animals as varied as fish, reptiles, and mammals, underscoring the benefits of live birth in some contexts. But across placental mammals, the structure never ceases to amaze. When you look at a mammalian heart, a lung, or even a brain, they look similar. The placenta, on the other hand, is wildly different across species. As reproductive immunologist Charlie Loke writes in his book about the placenta *Life's Vital Link*, which I had the pleasure of reviewing when it first came out, there is a seemingly infinite variety of placental shapes, sizes, and structures. No other organ varies so dramatically, something that has puzzled biologists for decades given how critically important it is for successful reproduction in placental mammals.

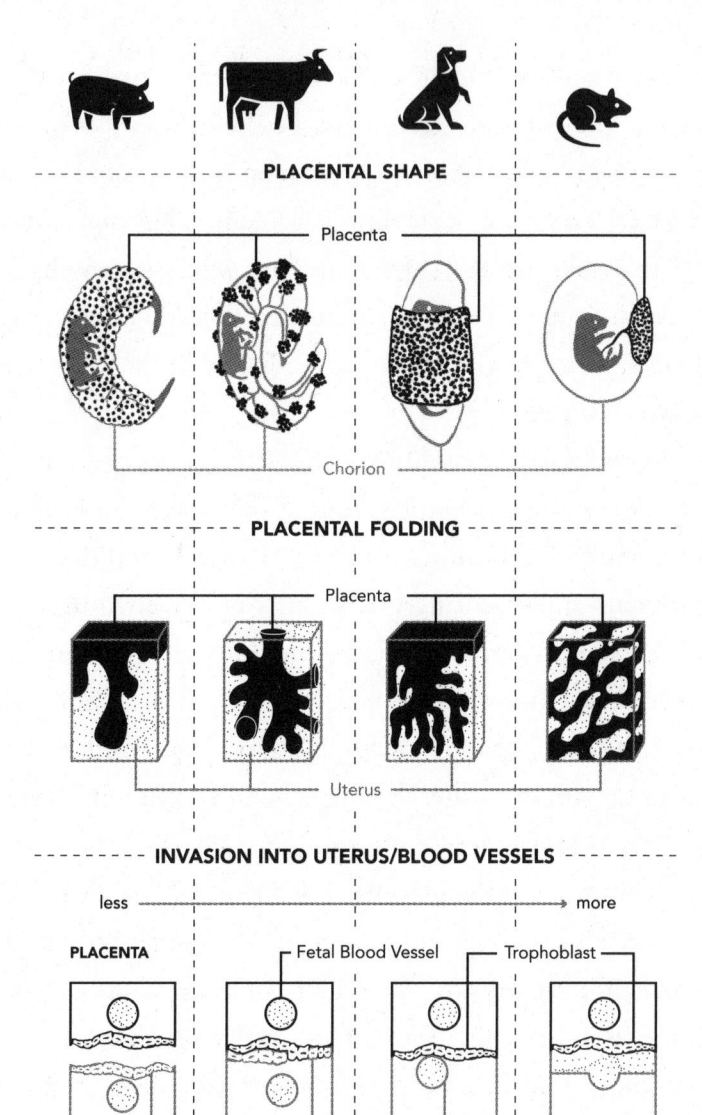

Different dimensions of placental variety: overall shape (top); folding of placental and uterine membranes (middle); invasion into the uterus and blood vessels (bottom). Across mammals, there is some mixing and matching of states in each dimension.

Just focusing on one aspect of the placenta—how deeply it invades the uterus—we think the earliest placental mammals had a moderately invasive placenta, thanks to those newly evolved invasive trophoblast cells. But then the descendants of this ancestor played around with the default mode in interesting and unexpected ways. In some groups, like the one to which cows and horses belong, the placenta became *less* invasive and is essentially barred from breaching the mother's uterine lining. The ability of these species to prevent placental invasion in the uterus may have had farther-reaching effects, specifically in how they manage cancer, as we'll discuss in chapter 8. At the other extreme is the ape placenta, including and especially our own, which became so invasive that trophoblast cells reach the muscle layers of the uterus. These are two ends of a continuum. And I'm describing only one of many placental dimensions, the others of which are also variable.

Why did this astonishing variation emerge? Mother-child conflict is likely a big part of the answer, which I'll dig into in the next section.

Mother versus fetus tug-of-war

Describing pregnancy as a battle or conflict is counterintuitive, and up to a point, the relationship between mother and fetus during pregnancy must be a cooperative one. Long gestations benefit mothers and babies, and there is no doubt that a complex placenta benefited placental mammals as a group. So once it evolved in early placental mammals, why was it subsequently changed so dramatically in different mammals? If it ain't broke, why fix it?

For years, biologists tried to correlate the type of placenta a mammal has with other aspects of its biology. Many thought a more invasive placenta evolved in species with big brains, enabling the more efficient transfer of nutrients required in brain development. But this is countered by the observation that some species, like dolphins, who are related to cows and horses and are endowed with huge brains and intelligence, have a noninvasive placenta. Clearly, it's possible to build a large brain without invading the uterus and tapping into the mother's blood supply. We do not yet know of any convincing connection between placental type and ecological or other biological factors that can explain why certain species require a certain type of placenta.

However, there is one evolutionary process that can explain the astonishing variety of placentas observed across mammals. We know that rapid evolution often occurs in situations in which two self-interested parties are interacting, such as a virus and a host. Remember from the conversation about mate choice in chapter 5 that the back-and-forth evolution of trait preferences in females and the traits in males explains the extreme and diverse forms of beauty in the animal world. Similarly, the back-and-forth evolution of placental tissues of the fetus and uterine tissues of the mother can explain why these tissues look so different across mammals.

In previous chapters, I discussed the tug-of-war over resources during pregnancy, first described by biologist David Haig, but now imagine a simultaneous tug-of-war happening between maternal and fetal genes across all placental mammals. These tugs-of-war can't possibly play out in the same way in each group or species, because evolutionary change requires mutations in DNA sequences, and mutations are rare and happen randomly. The differences in placentas that we observe today represent the different compromises reached in the conflict over nutrients. If we look at the ancestor of dolphins, horses, and cows mentioned earlier, it appears the mother gained the upper hand, and the placenta is forbidden from invading maternal tissues at all. When mothers in these species give birth, no blood is spilled because the placenta doesn't tap into blood vessels of the uterus.

At the other extreme, in apes and humans, it appears the fetus gained the upper hand, with placental cells invading even deeper into maternal blood vessels than in other species and a new subtype of placental cell that invades blood vessels in a novel way (from the outside of the vessel instead of from the inside). After the birth of my third son, I told my obstetric nurse that I wouldn't be able to "take it easy" with two other young children in tow, but then she graphically reminded me that childbirth leaves you with a huge, open, bleeding wound in your uterus. Not an ideal design! In another reality, humans probably could have made do with a noninvasive placenta, like that of dolphins, to support the development of a large-brained fetus. However, because of the specific mutations that occurred in our lineage compared to others, which influenced the particular way the tug-of-war played out, we're stuck with our inheritance.

Mother-child conflict helps to explain the incredible diversity of placental form across mammals, but if you're not convinced, there is plenty of other evidence for the existence of conflict in the womb. Let's start at the meeting point of maternal and placental tissues during a human pregnancy, which was often described by anatomists of the nineteenth and twentieth centuries using the imagery of a war zone. On the fetal side, you have placental trophoblast cells, which resemble invaders at the front line, advancing their attack deeper and deeper into the uterus. On the maternal side, you have decidual cells of the mother's uterine lining, which resemble a wall of shields. During their transformation into decidual cells, these uterine cells get larger and join tightly together to form a continuous block. In humans, we see just how invasive placental cells are and just how important decidual cells are in controlling placental invasion by observing what happens during pregnancies that take hold at sites without decidual cells. If you get pregnant in your fallopian tubes or on a scar from a previous C-section, both sites that lack a decidual transformation, invading tropho-blast cells are not restrained, potentially resulting in tubal or uterine rupture and maternal hemorrhage. These pregnancy complications are only experienced by humans (and other apes), an unfortunate consequence of how the conflict has played out in this group.

More evidence for the conflict is found when you look inside maternal and fetal cells at the front line and examine the molecules they're making. Fetal trophoblast cells produce enzymes that degrade the uterine lining. Decidual cells, on the other hand, make molecules that neutralize the effect of those enzymes. What exactly are trophoblast cells doing when they degrade the lining? They don't invade blindly; they specifically target the uterine arteries, and the goal is clear: to widen those arteries and replace the artery walls with their own cells, which increases blood flow to the placenta. More blood to the placenta means more opportunity for the fetus to take what it wants from the mother's blood. As a result of this remodeling, a human mother can't reduce blood flow to the placenta (without also reducing it to her own organs). That said, maternal countermeasures were likely put in place. The uterine arteries in species with invasive placentas have an unusual spiral shape, which is thought to increase *resistance* of blood flow to the placenta. It's like the flow of water out of a garden house that is coiled up versus straightened out—there is more resistance in the coiled hose, and the water dribbles instead of gushes out.

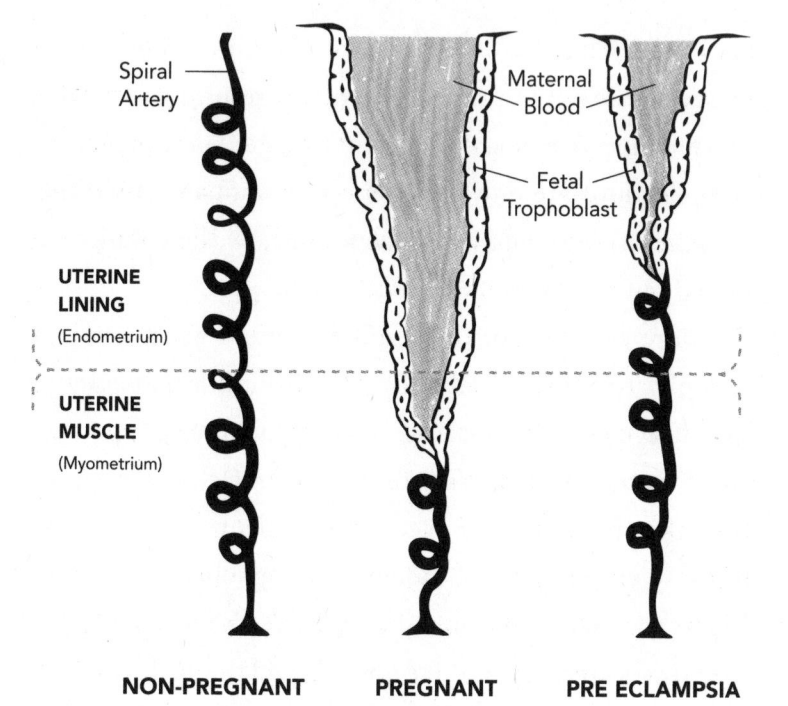

NON-PREGNANT   PREGNANT   PRE ECLAMPSIA

**Spiral artery transformation of the uterus
during human pregnancy.**

Illustration based on figure from *Life's Vital Link* by Y. W. Loke

In some human pregnancies, the spiral arteries don't get suffi-ciently widened for reasons that aren't clear—possibly the placenta isn't invasive enough, or the endometrium is too restrictive. The result is that the placenta, and ultimately the fetus, doesn't get enough oxygen or nutrients, which can lead to a condition called preeclampsia or the more severe eclampsia. If you've watched *Downton Abbey*, you'll be familiar with Sybil's tragic death during childbirth because of preeclampsia/eclampsia, which is the most common cause of maternal death in humans. In cases of

in over-the-counter pregnancy tests. This hormone doesn't signal to fetal or placental tissues but rather signals to the mother's ovary to keep making progesterone. It binds to the receptor of a different hormone, luteinizing hormone (LH), which is made by the mother's brain and signals to the ovarian follicle to make progesterone, which is required to continue a pregnancy. So in effect, the production of hCG by the placenta transferred some control of continuing a pregnancy over to the fetus. Sneaky little bugger! However, in the ongoing conflict, mothers may have started using hCG levels as a signal of embryo quality, evolving ways to reject the ones that weren't making enough. The placenta may have responded by making more hCG. Back and forth, back and forth. Pregnancy hormones like hCG are found at bizarrely high levels, *much higher* than necessary, which is precisely what the conflict hypothesis predicts should happen. High hormone levels are the equivalent of a shouting match. You shout to be heard by the person with whom you're arguing, but because the other person is shouting too, it's hard to hear (or believe) anything and you've both wasted a lot of energy in the process. One possible side effect of such high levels of hCG (and other placental hormones) in the first trimester is morning sickness, a phenomenon of pregnancy that is still not well understood.

Another set of hormones expressed at unusually high levels during pregnancy are those that control glucose levels in the mother's blood. After a meal, food is digested, and one type of energy that is released from the food—glucose—travels in the blood to be taken in by cells that need the energy. During pregnancy, this energy needs to be shared by the mother and the fetus. The placenta dumps hormones into the mother's blood that keep her blood glucose elevated after a meal, which gives the fetus a longer time to grab glucose from her blood at the placenta. One of these hormones is placental lactogen, which is the hormone in primates that is produced at the highest concentrations. The production of placental lactogen and other similar hormones can be viewed as fetal tugs in the tug-of-war over energy. Mothers may have responded by making more insulin, which is the hormone that moves glucose from our blood into our cells, thus reducing blood sugar levels. A tug-of-war ensued over evolutionary time until levels of these hormones made by both parties reached extremely high levels in humans, much higher than observed during nonpregnant conditions. The placenta even makes enzymes that degrade insulin—another fetal tug in the struggle over nutrients. A pregnant mother makes peak levels of insulin during the third trimester, which paradoxically coincides with the time that her blood sugar is the *most* elevated. So maternal insulin seems to be less effective during pregnancy because of the counteracting action of hormones and other proteins made by the placenta. In cases of gestational diabetes, mothers are unable to mount an effective enough response against these placental hormones, resulting in high blood sugar, which can lead to high blood pressure and preeclampsia as well as an excessively large fetus and difficult birth.

Another arena for conflict (and cooperation) during pregnancy is the mother's immune system. The job of the immune system is to recognize and destroy outside invaders of the body, such as viruses and bacteria, and the vertebrate immune system evolved sophisticated ways to target anything that looks different from the body's own cells. Since the physical identity of our cells is determined by our genes, a long-standing puzzle is why the mother's immune system doesn't reject a fetus that is not genetically identical to her the way it rejects viruses, bacteria, and, in cases of organ transplants, organs from another person. Clearly, flat-out rejection is not good for mother or fetus, and the immune dialogue during pregnancy reflects both the shared and divergent interests of mother and fetus.

In early stages of pregnancy in placental mammals, a ubiquitous type of white blood cell that travels to wounds after injury is *prevented* by the mother's uterus from traveling to the implantation site. If the mother didn't thwart these immune cells, the embryo would be rejected, highlighting the cooperative nature of the maternal immune system. But mothers also recruit into the uterus a novel type of white blood cell with special features that allow them to control placental invasion. These special white blood cells don't reject invading placental cells, but they do help keep them in check. On the fetal side, different populations of placental cells have some unusual features that help them evade the mother's immune system or convince it to cooperate (similar to tactics used by viruses and bacteria to evade immune defenses). There is a division of labor among placental cells, and depending on their precise function, some remain invisible to the mother's immune system, while others express novel proteins on their cell surface that allow them safe passage. In apes and humans, which have the most invasive type of placenta, the dialogue has escalated, with both the mother's immune system and the placenta making novel immune proteins that mediate the deep invasion. Since these immune proteins have more classic functions outside the uterus, it is unclear whether they initially evolved to fight infections in apes or for pregnancy-related functions. A full discussion of the immunology of pregnancy requires a whole book or at least a whole chapter, which is beyond the scope here. If you're interested, I recommend the volume I mentioned earlier, *Life's Vital Link*.

The most convincing (and surreal) evidence for conflict in the womb—genomic imprinting—falls on a different dimension from the examples I've used above. In all these examples, I discussed moves by the fetus (via its henchman, the placenta) that are matched by countermoves by the mother. The conflict is between genes in two different individuals, mother and fetus. Since it's the father's genes in the fetus that differ from the mother's, the conflict is really between maternal and paternal genes over nutrient transfer to the fetus. But there's an additional dimension to this battle, happening completely *within* the fetus and its placenta. This battle is between genes in the fetus/placenta inherited from the mother and those inherited from the father. This may sound bizarre, so to explain it, I need to go back to some of the basics from chapter 1 on early human development.

Recall that human eggs have twenty-three chromosomes, sperm have twenty-three chromosomes, and the full complement of forty-six is reached when egg meets sperm. In most cases, genes in the developing baby have no clue whether they came from mom's twenty-three chromosomes or dad's—a gene is a gene, whether it came from mother or father. But in some cases, they do have a clue. A whole new system of inheritance called genomic imprinting evolved in mammals, specifically in the ancestor of marsupials and placental mammals. In genomic imprinting, fetal genes retain memory of which parent they came from because of special biochemical marks deposited on DNA in eggs and sperm. These marks often prevent the gene(s) they mark from being active in the baby. Not many genes are marked—on the order of two hundred in humans and mice, out of perhaps a total of twenty thousand—so being marked is rare and special. But intriguingly, many that are marked play a role in placental and fetal development.

Even more intriguingly, imprinting affects pregnancy in ways that are consistent with mother-child conflict. The imprinting marks deposited by dad push for bigger placentas and bigger babies, and those deposited by mom favor restraint. There is an imprinted gene called insulin-like growth factor 2 (*IGF2*), which helps trophoblast cells invade the uterine lining. Only the father's copy of *IGF2* is active; the mother's is silent. Conversely, there is another gene, *IGF2R*, that codes for a protein that binds to *IGF2* and prevents it from working. Only the mother's copy is active. Many imprinted genes behave according to this pattern, which I think is incredibly cool and provides support for the idea that genes from mothers and fathers have been in conflict over resources to children during pregnancy. I'll discuss genomic imprinting in the next section and in more detail in the next chapter.

There are multiple lines of evidence suggesting that a tug-of-war over maternal resources is responsible for the surprising way pregnancy evolved in humans and other mammals. Genes from the mother and fetus agree on the big picture—a successful pregnancy—but they often disagree about the details, and the "bickering" over these details has resulted in some unusual aspects of pregnancy. I've been using the tug-of-war analogy to describe how pregnancy evolved over evolutionary timescales, but we can use the same analogy to talk about how our individual pregnancies unfold. In most pregnancies, mother and baby pull equally hard on the rope, the way they've evolved to do, and no one falls down. But if the balance of maternal or fetal factors is slightly off, a woman or her baby may experience one of the many pregnancy complications that afflict our species. As David Haig has argued, the evolutionary contest between mother and fetus for nutrients has created instabilities in many aspects of the physiology of pregnancy.

The original parasites of pregnancy

In training to be a scientist, one often starts with some general interests that become more specific over time. As a master's student at NYU, I was vaguely interested in human evolution. As I completed my master's and began a PhD program at Yale, I became drawn to questions about the kinds of changes in DNA that drive changes in body parts and physiology and behavior (in humans or more broadly), DNA changes such as those that allowed an egg-laying mammal to start growing babies inside its body. How do such radical changes evolve?

I happened to join a lab at Yale that was studying some strange DNA elements—transposons—and their role in the evolution of pregnancy. To understand the part these elements have played in the pregnancy story, let's first take a step back and talk about the structure and function of our genome, which is all the DNA that is present in each of our cells. We've discussed the forty-six chromosomes that humans have in each cell (except for eggs and sperm, which have twenty-three). Each chromosome is one long DNA molecule, constructed with building blocks called nucleotides, that gets coiled up with some packaging proteins. If you scan from one end of a chromosome to the other, you'll find genes along the DNA. Genes are the blueprints for building proteins, which are the workhorses of the cell. When you think about DNA, perhaps the first thing that comes to mind is genes, but only about 1 percent of all the nucleotides that make up our DNA are found in genes. What's going on with the other 99 percent?

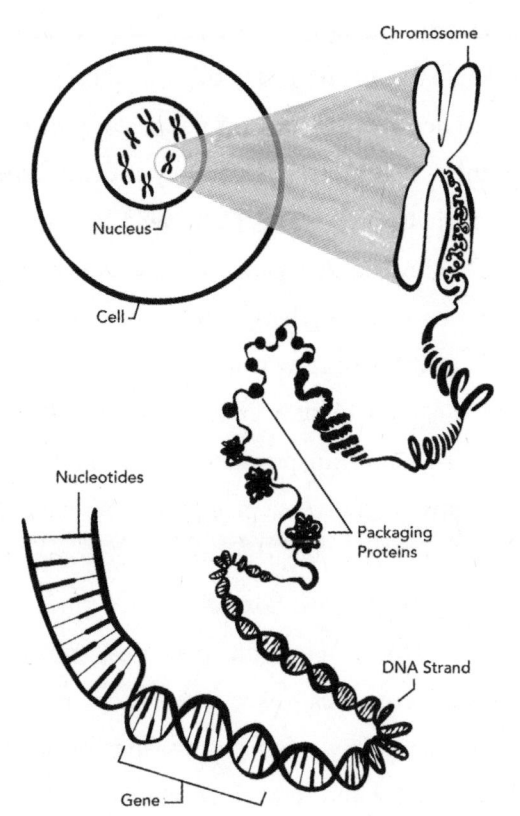

DNA, chromosomes, and genes.

Illustration based on original by National Human Genome Research Institute.

It used to be called "junk"—possibly useful but not entirely clear how. We now know that the DNA regions around genes are involved in gene regulation—in other words, controlling when, where, how much, and how long a gene is expressed, or active. Think of a dimmer switch controlling a lamp in your living room—without the ability to turn the lamp on, off, up, or down, the lamp is not very useful. Genes are like lamps, useless without some way to control their activity. The dimmer switches located around genes contain many short DNA sequences that are recognized by specific proteins that control the activity of the gene. Gene switches are a necessary feature of life, allowing an organism to use a limited number of genes in *many* contexts, such as in different cell types, different stages of life, or different environments. That said, these gene switches don't explain all or even most of what was thought of as junk. This is where transposons come into the story.

Transposons are a diverse group of genetic elements that together compose over *half* the human genome. They have some surprising properties. Sometimes described as jumping genes, selfish DNA, or genomic parasites, transposons can copy and paste themselves—over and over again—into a host's genome. Like parasitic worms that invade your gut and feed off your blood to live and reproduce, transposons invade your genome and use your cell machinery to reproduce themselves *within* your genome. As one of the recurring themes in this book is genetic conflict that exists between two interacting parties, transposons are a perfect topic to include in our discussion of conflict. In this case, the two battling parties are the transposon and the host genome.

Successful transposons are those that manage to spread abundantly around a host genome. If they jump around in eggs or sperm, the new copies (and old ones) are transmitted to the next generation of host. But they can be harmful to the host if they jump into important regions, like those that contain genes or gene switches, so hosts have evolved methods to shut them down. Interestingly, one of these methods—putting biochemical marks on transposon DNA that prevent it from being active—bears a striking resemblance to that used during genomic imprinting mentioned in the last section, which enables parents to shut off certain genes in eggs and sperm. In fact, some have argued that genomic imprinting evolved from this host-defense strategy to shut down transposons, which were particularly active in the ancestor of marsupials and placental mammals, the *same* groups that have genomic imprinting. But as hosts evolved ways to shut transposons down, transposons evolved ways to evade these defenses in an ongoing tug-of-war between transposon and host genome. When you look at the human genome today (or any animal or plant genome for that matter), you observe the end result of many such conflicts from the evolutionary past. Our genomes are littered with different types of transposons that spread around the genome at different points in time, with varying degrees of success. Most were eventually shut down, and only a small number still jump around our genome today.

These conflict dynamics are fascinating in their own right, but to the evolutionary biologist interested in *how* bodies evolve at the level of DNA, one of the most intriguing things about transposons is that they provide lots of new DNA sequence to a host genome in a short amount of time. That new sequence could be an engine for evolutionary change in the host. After a transposon gets shut down, it continues to be part of the host genome, usually in multiple copies. These defunct transposons compose much of the junk DNA that I mentioned earlier. But instead of just sitting in our genome doing nothing, individual or multiple copies of the same element can be recruited to do useful things for the host. Transposons can supply a genome with new genes and gene switches, which—depending on if, when, where, and how many are recruited—can lead to big changes in the host. This is exactly what we think happened during the origin and subsequent evolution of the placenta in mammals.

One feature of transposons that makes them attractive candidates for driving the evolution of the placenta, which is so variable across mammals, is that their activity is quite specific to certain lineages. In other words, one element might be active in humans, a different one active in horses, and yet another in bats, each one spreading a *different* DNA sequence around the genome of that species. Even if the same transposon is active in two different species, it will jump into different places in their genomes, which could influence if and how each copy is recruited. These differences in transposon activity in different mammals could explain how the placenta/uterus has diverged so dramatically in this group. What is the support for this idea?

In the fetus, there is ample evidence that a class of transposon called the endogenous retrovirus (ERV) was recruited in the placenta, not just once but multiple times in different mammals. ERVs are a special type of transposon that descends from infectious viruses that integrated into the host genome. Like many transposons, ERVs have their own genes, coding for some proteins that help cells fuse together, some that allow cells to evade the immune system, and some that enable cells to invade tissues. Do these functions sound familiar? They are all properties of the invasive placenta. We know that distinct ERV transposons were recruited in different mammals, including in primates, rodents, and sheep, each playing an active role in placental development and function. We also hypothesize that ERVs were recruited in the ancestor of all placental mammals, enabling the original placental invasion into uterine tissues. In addition to ERVs, other types of transposons have been recruited in the placenta, many acting as gene switches, allowing genes that previously weren't active in the placenta to be active.

In the mother, there is also evidence that transposons were recruited in uterine tissues, driving some of the changes that were necessary for pregnancy to evolve. In a project led by Vinny Lynch, a colleague from Yale who is now at the University at Buffalo, we showed that transposons that were active in the genome of early placental mammals spread around gene switches important for the transformation of uterine cells into decidual cells. In other words, the origin of the decidual cell in the first placental mammals involved the recruitment of *thousands* of transposons in the uterine lining. Evolving a new cell type is no small feat—different cell types have vastly different jobs to do, using a different set of genes and gene switches to do those jobs. The recruitment of these transposons enabled uterine cells to do a new set of jobs critical for pregnancy. Many types of transposons and transposon copies were recruited, but one thing that they have in common is short DNA sequences that respond to progesterone. Progesterone is the hormone critical for decidual transformation, decidual function, and maintenance of pregnancy. So the transposon-driven buildup in the genome of gene switches that respond to progesterone was a key event during the evolution of pregnancy in mammals.

These studies and many others have shown how important transposon recruitments were for the origin and continued evolution of pregnancy in mammals, driving changes in maternal and fetal tissues that interact so intimately during pregnancy. This fascinating body of work tells us that one route to major body changes (like those that allowed mammals to start growing babies inside their bodies) and smaller changes (like tweaks to the original placenta in different mammals) is the recruitment of transposons. Transposons have played a huge part in our pregnancy story.

———

To wrap up this chapter on pregnancy, let's zoom out from the microscopic back to the macroscopic players of pregnancy: the mother and fetus. We've all heard the developing fetus described as a parasite, perhaps by a woman in her third trimester who is beyond fed up with being pregnant. The fetus lives in and feeds off its mother, just as a parasite does its host. Whether or not you agree with this unsentimental view of pregnancy, I find it fascinating that the very thing that enabled this parasitic relationship—the transposons I just described— are parasites themselves, exploiting our cells for their own selfish purposes. But then fetus and mother exploited these parasites for *their* purposes, their own conflict (and coopera- tion) in the womb. This parasite within a parasite brings to mind the "strange loops" discussed in Douglas Hofstadter's book *Gödel, Escher, Bach*, a phenomenon in which you move through levels of a system but somehow end up back where you started. Like M. C. Escher's lithograph *Drawing Hands*, or Russian nesting dolls, or dreams in which you're dreaming about dreaming (as in the movie *Inception*), it is mind-boggling to think about the conflicts at one level (transposon versus host genome) that have enabled conflicts at another (maternal versus fetal genes).

Continuing with the theme of conflict, in the next chapter we'll discuss how the genetic conflicts during pregnancy do not end at birth but extend into childhood and beyond.

CHAPTER 7

The Conflicted Mother

I was struck with two contradictory feelings after delivering my first son, neither of which was the flood of maternal elation that I expected. The immediate feeling was one of lightness, a nine-pound load having literally been lifted away from my body. But in the hours after my son's birth, a new kind of weight descended on my shoulders. The feeling reminded me of the Greek myth in which Heracles briefly holds up the heavens while Atlas picks an apple from the garden of the gods. The overwhelming weight of responsibility I felt on the first night of my son's life is how I imagine Heracles felt when the weight of the sky was suddenly transferred onto his shoulders. While my husband snoozed obliviously in the hospital chair next to us (watching someone give birth must be exhausting!), I had to figure out how to hold my son, soothe him, and, most daunting of all, feed him.

When Atlas returns with the apple in the Greek myth, he and Heracles each try to trick the other into holding up the heavens.

But Heracles outsmarts Atlas and leaves for his next adventure. Although I too was overwhelmed, in my case with the responsibility of a new child, I continued bearing the weight. Over the next few weeks, I learned how to calm my son, bathe him, and feed him, all while being severely sleep deprived. You could say I made the willing choice to fall for my son's tricks, of which, as you'll learn in this chapter if you don't know already, children have many. And since we know from previous chapters that a child's subversive ways stem from dad, perhaps I should say I fell for my snoozing and carefree husband's tricks!

Those feelings of contradiction in the moments after my son's birth never went away. After his delivery, I was both lighter and heavier. During the years of breastfeeding my infants, I felt euphoric at times and enslaved at others. When I dropped my babies off at day care in the mornings, I was at once liberated and anguished. More generally—then and now—I'm consumed with an obsessive love for my children interspersed with the frequent desire to escape to a remote island where no one can bother me. If I had to choose one word to describe my experience of motherhood, it would be *ambivalent* or *conflicted*. This admission feels like mommy sacrilege—good mothers are supposed to sacrifice themselves for their kids and love every minute, or so the world tells us. But as I'll discuss in this chapter, there is an ancient biological basis for maternal emotions and decision-making that helps explain the feelings of contradiction that human mothers do experience, even if we're loath to admit it.

The evolution of a relationship as intimate as that between a mother and child, whether fetus, infant, or tween, is a *co-evolutionary* process. Traits and behaviors that emerge in one party don't evolve in a vacuum, and often the two parties evolve back and forth and in response to one another. As we've talked about in previous chapters, when individuals must interact to achieve some common goal—such as successful transfer of sperm from male to female or successful pregnancy between mother and fetus—often the genes of interacting parties agree on the overall goal, but they disagree on how to achieve it. Both male and female "want" fertilization to happen; both mother and fetus want to make it through pregnancy; both mother and newborn want the transfer of milk. But the two parties may disagree on the details of the relationship—such as with whom, when, and how long—because their evolutionary interests are not identical.

In the following pages, we'll discuss the evolutionary agendas of mothers and their children (and other vested parties), which only partially overlap. The genetic negotiations between mothers and children—in our mammalian, primate, and human ancestors—helped shape the range of emotions and behaviors that mothers and their children exhibit today. Human mothers are unique in the animal world for having the intellectual and technical capacity to override these maternal "instincts" (thinking of the many women who choose not to have children). But as you'll learn, these instincts—which are remarkably flexible in the animal world—remain with us today, influencing our emotions and experiences of motherhood.

What a child wants, what a child needs

One subject that consumes many first-time mothers is sleep. I'm someone who likes my eight hours of sleep, and while I was pregnant with my first son, I lost sleep over just the thought of losing sleep, of being woken up repeatedly in the night by my future needy newborn. After my son was born, I wearily made it through our first few months together but continued to be obsessed with sleep. The morning conversations I had with other mothers at day care drop-off were invariably about sleep. I might have shared with another mom that my four-month-old slept a five-hour stretch the night before, or to avoid the evil eye after a particularly good night, I might have kept the details vague. I agonized about if, when, and how I should sleep train my son so that I could get more sleep. My preoccupation with sleep continued with my other babies and continues to this day, because older kids wake their parents up too. Why are children so needy, not just at night but twenty-four seven? What do they need, exactly?

In primatologist Sarah Blaffer Hrdy's insightful book about motherhood, *Mother Nature*, she puts the needs of a human child in a mammalian and primatological context. In contrast to many mammals who produce large litters of offspring, monkey and ape mothers typically have one baby at a time. These singleton babies require substantial investment by the mother—a chimpanzee isn't weaned until it is about five, and the mother is the sole provider of milk and care until weaning. Monkey and ape babies spend much of their early lives literally hanging on a thread—in their case, the fur of their mother. The Moro or startle reflex in human babies—flinging the arms and legs out and pulling them back in quickly—is likely a relic from our primate past when clinging was critical for survival. Clinging to mom brings protection, warmth, and a dependable supply of milk, and importantly, in the first few weeks of the baby's life, it induces hormonal and brain changes in the mother that commit her to the relationship (more on this later). Primate mothers make very dilute milk because their babies cling to and nurse from them all day and night, in contrast to some mammals who leave their babies in a nest or den for extended periods of time so need to make more concentrated milk. Some experiments from the '60s on baby rhesus monkeys showed that it's not only or even mainly milk that babies are after. These monkeys were separated from their mothers and offered two substitutes—one surrogate made from wire and holding a baby bottle, and one covered in a soft and furry fabric but with no bottle. Contrary to what the researchers expected—that the monkeys would spend all their time on the surrogate with the bottle—the monkeys spent only enough time on those to take their milk and spent most of their time clinging to the soft ones. The soft surrogates provided something to these baby monkeys that milk alone didn't satisfy.

Because most monkeys and apes produce only one baby at a time and those babies are spaced so far apart, nonhuman primate mothers are particularly devoted moms (compared to other mammals and surprisingly, as you'll learn, to humans). Hrdy describes some heart-wrenching scenes of monkey and chimp moms carrying around the carcasses of their dead infants for days. Once they've committed, these mothers care for each baby, regardless of its physical state. This behavior likely served past primate mothers well—their unquestioning devotion kept some babies alive that would have otherwise died. For a chimp that only produces one baby every six years or so, that's a big evolutionary deal.

Most mammalian mothers need to be more ruthless in calculating if and how much they will invest in each child. In many species, such as the coypu rodent, the pregnant mother's body will spontaneously abort a litter if the litter is too small or doesn't contain the ideal ratio of males to females. After birth, house mice will push aside the smallest of a large litter and focus all their milk and care on the stronger ones. Golden hamster moms will *eat* the weakest of a large litter to adjust litter size to one they can manage and recoup some of their investment in the process. Syrian hamster moms will eat specific pups to adjust the ratio of males to females in their litters. Even primate moms, especially those who have more than one baby at once and when conditions are far from ideal, are forced to make similar decisions. One example is tamarin monkeys, which often give birth to twins or triplets. Tamarins are rare primates that breed cooperatively, which means that many group members contribute to raising offspring, not just the mother. If a tamarin mother gives birth to multiples but doesn't have family around to help rear them, the mother will abandon the babies within their first few days of life. She instinctively knows that committing to these babies without any help is a futile endeavor that will delay or impede her future reproductive successes.

As Hrdy explains, the commitment of all mammalian mothers, including primate and human ones, is highly contingent on circumstances. The brains and bodies of female mammals are wired to evaluate trade-offs, such as continuing to grow or stopping growth to reproduce, investing in a child now or waiting until conditions improve, investing in fewer or more children, investing in this child or that one. Maternal decisions are never automatic but depend on the context, such as how old a female is, how healthy she is, how much food is available, how many other children she has, and how much help she has. A mother's big picture necessarily includes all these variables, because they influence how successful she will be as a mother in the long term. Flexible maternal decision-making enabled mammalian mothers to make the most of unpredictable and often difficult circumstances.

What about humans, who have a mammalian and a primate legacy? Humans are unique in many respects, possessing a suite of traits that allowed our species to dominate the globe both physically and technologically over the last one hundred thousand years. One key difference between humans and our closest primate relatives, chimpanzees, is that our babies are weaned much earlier, at two to three years in populations that don't use birth control compared to five years in chimps. Human children also develop more slowly than other primates, reaching reproductive maturity and independence many years after chimp children do. So not only do we wean our babies earlier, but these babies depend on us for much longer. In contrast, once a chimp weans, it is responsible for feeding itself. Because the act of nursing is a natural birth control, preventing a mammalian mother from ovulating, earlier weaning in our human ancestors translated to babies being born at much shorter intervals. Although most women have one baby at a time like our primate cousins, the traits of early weaning and longer childhoods in effect left human mothers caring for "litters" of dependent children of different ages. The faster reproductive rate of humans relative to that of other primates is one of a few reasons we have come to dominate the globe, but it would have put our human ancestors in a tough spot. How could they possibly care for so many children at once?

As we'll discuss later in this chapter, one critical development in humans that enabled this faster reproductive rate was the help provided by other family members, including fathers, grandparents, and older children, in raising young children. Humans became cooperative breeders, a rare trait in primates and mammals more generally. Nevertheless, human mothers also had to reemploy some of the more calculating maternal behaviors observed in other mammals, like hamsters, because help was never a guarantee. In *Mother Nature*, Hrdy takes her reader on an eye-opening journey through human history, which is filled with stories of infant abandonment, infanticide, and sex-specific preferences that have occurred on a massive scale. One such story is that on the order of *millions* of babies in Europe during the Renaissance and the Enlightenment were left in foundling homes—orphanages for babies—where they had an exceedingly poor chance of survival. Before baby formula was invented, an abandoned infant required a wet nurse for survival, a costly and rare service. Females and later-born children have fared particularly poorly all over the world. In cultures in which one sex is valued more than the other (usually males), parents routinely abandon or kill female newborns and—now with ultrasound technology—abort female embryos, leading to highly skewed sex ratios. One story from the early '90s that stuck with me is of a Pakistani mother who birthed girl and boy twins. The boy stayed with the mother and nursed, and the girl was sent to the in-laws and was bottle-fed. Five months later, the trio was reunited. A photo was taken of the mother nursing her plump, healthy boy on one side and bottle-feeding her tiny, emaciated daughter on the other. The daughter died of malnutrition not long after the photo was taken. This story hits a nerve because I, like this unfortunate Pakistani girl, have a twin brother, and in another time or place, my fate might have been like hers.

These stories and statistics have been used to argue that maternal instincts are a cultural invention, with no biological basis. If women had mothering instincts, how could so many of them throughout human history commit such heartless acts? Hrdy argues otherwise, likening the decisions that human mothers make to those that all mammalian mothers make regularly. First, human mothers most likely to abandon their babies are those without the means or help to care for their children, like those tamarin monkeys without family nearby. Second, human mothers in the past (and still today) lived with other group members who might have been more powerful than the mother, pushing their own agendas, such as the Pakistani woman's husband and family deciding the son was more valuable than his sister. In many primate species (thankfully, not ours), the agenda of males includes the practice of infanticide—killing a female's current breastfeeding babies, whom the male is guessing aren't his, so he can more quickly impregnate her for his own gain. Because the males in these species are so much larger, stronger, and better weaponized than the females, there isn't much a female can do but acquiesce (but stay tuned on this point).

Another important point Hrdy makes is that humans are like the primate mothers mentioned earlier, whose hormone levels and brains change once their babies start clinging and nursing, making them much less likely to abandon their infants. Prolactin is an ancient hormone that rises in parents (both male and female) of many species engaged in nurturing behaviors, including in seahorses, birds, and mammals. Among its many roles in mammals, prolactin promotes lactation, the function from which the hormone derives its name. Nipple stimulation signals to the mother's brain to release prolactin, which in turn acts on the milk-producing cells of the breast. Prolactin also suppresses ovulation, acting as a natural birth control in lactating mammals. Oxytocin is released when a mother is breastfeeding and cuddling with her children (and during many other activities across a range of mammals, such as grooming, sexual activity, and social interactions), encouraging mothers to continue engaging in these activities. Oxytocin is a natural opiate, activating dopamine-producing cells in the brain that are central players in the brain's reward system that we discussed in previous chapters. Brain changes that happen in new mothers include the growth and increased connectivity of certain parts of the brain, including those involved in making and responding to oxytocin. Breastfeeding and physical contact with our children literally change the structure of our brains to reinforce maternal behaviors. As Hrdy has argued, human maternal commitment might not kick-start immediately after birth—most abandonments happen in the first few days of life—but once the commitment is made, etched in the hormonal milieu and brain circuits of new mothers, it is exceedingly difficult for a mother to retract it completely.

So back to the question posed earlier: *What do human children need, exactly?* Imagine the world of our human ancestors in the last one hundred thousand years and even longer ago, in which the availability of food and childcare was unpredictable, animal predators lurked, and human predators lurked too. Even though human males don't commit infanticide in the calculated way of some primate males, group and tribal conflict took a disproportionate toll on young children and mothers. Our ancestral mothers were certainly put in impossible situations. They did not have the option of birth control as we do today. But they did have the capacity to make decisions about keeping or abandoning a child depending on their circumstances and about distributing care among current children in a way that furthered their own agenda. The haunting scene from the film *Sophie's Choice*, in which Sophie is forced to choose which of her two children will be sent to a Nazi gas chamber and which will be sent to a children's camp, illustrates the types of decisions mothers have been forced to make for millennia. In agony, Sophie chooses her son to go to the camp because she believes he has a better chance of surviving; if she chooses her daughter, she'll end up with two dead children.

Back to the harsh and unpredictable world of our ancestors, in which mothers were constantly making difficult decisions. Children needed maternal commitment above everything else. A committed mother meant food, protection, and warmth. A committed mother was a child's ticket to survival. Primate babies want commitment too, but comparatively, human babies are at a disadvantage. They have no fur to cling to, and at birth, human brains aren't developed enough to even think about clinging (this is one of the trade-offs of a large brain and bipedality, which limits the size of the birth canal and thus brain size at birth). Human babies had to get creative about how to secure maternal commitment since hanging on was not an option, and older children had to continue reminding mothers of that commitment (Sophie's daughter's screams while being ripped out of her mother's arms were an attempt to remind Sophie of her commitment). Many of the traits and behaviors of our children today are relics of an ancient time when they made the difference between life and death. Children who adopted these behaviors had a better chance of staying alive and passing those traits on to the next generation. I talk about some of this child ingenuity in the next section.

Every (kid) trick in the book

Living in a house with four playful kids and a jokester husband, I've come to appreciate the range of tricks one can play on others, some harmless, some annoying, some bordering on dangerous. There's the innocent "fake egg for breakfast" trick, in which you tell someone you're going to fry them an egg and present them instead with a half peach surrounded by yogurt that looks a lot like an egg if you do it right. Both parties win with this one—you get a laugh when your "victim" takes their first bite, and your victim gets a healthy breakfast and hopefully a laugh as well. On the nastier end of the prank spectrum, you could put laxative powder in your housemate's coffee machine or itching powder on the toilet paper in their bathroom. I suggest a combination of both for folks you really have it in for.

Just as there is a range of tricks available to the enterprising prankster, babies and children have an arsenal of evolved "tricks" they use on their moms to get what they want, some less harmful than others. Birth signals a major turning point in the genetic conflict of pregnancy that we discussed in the last chapter. The conflict does not end at birth, but the balance of power shifts. In species with invasive placentas, the fetus has the upper hand during most of pregnancy; it has direct access to the maternal blood supply, which allows it to manipulate mom with hormones and other molecules. But once that baby is born, the power shifts back to mom. She can abandon the baby if she must, and she can withhold milk if she so chooses. Biologist David Haig has hypothesized that one of the first tricks an infant plays on its mother is staying in the womb longer than is ideal for the mother. He reasons that the length of pregnancy in humans is the outcome of an ongoing tug-of-war, with paternal genes pushing for longer gestations (and consequently bigger babies) and maternal genes pushing for shorter ones.

We don't understand what (or who) triggers birth in humans, information that would be extremely valuable in the prevention of preterm birth. In many mammals, the mother controls progesterone production during pregnancy, and she controls the rapid withdrawal of progesterone to initiate labor. But in humans (and other apes), the baby's placenta takes over progesterone production, and progesterone levels never drop during pregnancy, even at the end. This has been interpreted as a fetal move to prolong pregnancy. Mothers in these species have likely evolved another way to initiate labor, but we don't know what it is. While the current evidence that conflict has influenced gestation length in humans is somewhat limited (we'll discuss it more later), there is no doubt that human babies are born too large, with massive heads and shoulders. Anyone who has birthed a child has cursed the stupidity of the design, of our huge babies having to squeeze through an opening that is much too small. No other mammal suffers as we do, except perhaps the hyena mother, which must deliver its babies through a pseudo penis.

When you compare a human to a chimpanzee or monkey baby, one thing is immediately clear—our newborns are fat, making primate babies look scrawny in comparison, even those of the gargantuan gorilla. Yes, human heads are large at birth, but about 15 percent of our weight at birth comes from fat tissue, much of which is distributed around the shoulders, contributing to a tight fit of the shoulders during birth in addition to the head. Many anthropologists have chimed in on why human babies evolved to be born with so much fat. One idea is that fat provided insulation from the cold, another that it was insurance against emergencies, like the temporary disappearance or even death of a mother. Since these are problems many primates face but none other than humans evolved such a large stockpile of fat before birth, these

babies to feed. Like coots, some monkey babies are born with flashy colors, such as the bright orange coat of the silvered leaf monkey baby, which fades to a dull gray as the individual grows up. The stunning color likely serves to attract attention, hopefully not from a lurking predator but instead from the mother or a relative being enlisted to help care for the baby.

Human babies, in particular, would have benefited from flashy advertisements that say "Choose me!" given their mothers' large "litter" sizes and an increased intellectual capacity to make choices between children, like Sophie's choice of son over daughter. As birth weight is a simple predictor of a newborn's health and survival prospects—a connection appreciated by many cultures around the world and supported by many scientific studies—the evolutionary process by which human babies became fatter may have started with mothers choosing to invest in slightly fatter babies over skinny ones. But once that preference was established, Hrdy hypothesizes that Fisherian runaway selection ensued (see chapter 5). Babies started putting on even more fat in the last trimester, and mothers developed an even stronger attraction to fat babies, to the point that after having four babies of my own, I still feel possessed to hold any chubby infant with whom I cross paths. Beyond just baby fat, Hrdy posits that many infant traits that have us grabbing at other people's babies, such as those irresistible smiles, coos, and giggles, evolved as baby advertisements. In contrast to the long-held view that babies are just passive vessels into which mothers actively pour their investment,

Hrdy argues that "human infants have been selected to be activists and salesmen, agents negotiating their own survival."[1]

Beyond being plump at birth, more obvious newborn tricks include rooting for the breast and suckling, behaviors that amazed me in my own newborns given how helpless they were otherwise. With my fourth child, I had the experience and confidence to run my own little experiment—I placed my just-born daughter on my belly and watched her worm her way up to my breast. Not much nutrition or water is gained in these early breastfeeding sessions—it takes a couple of days for a mother's milk to start flowing. But as we've been discussing, nipple stimulation and physical contact elicit hormonal changes in the mother that make her feel good, encouraging her to continue engaging in these behaviors and committing her to the relationship. Babies in the past who had strong rooting and suckling reflexes certainly triumphed over those who didn't.

Once maternal commitment is established after birth, what is the next big transition in a child's life that would benefit from some tricks? Weaning time. Indeed, the first theorist to suggest that children were not passive vessels but active agents with their own agendas was Robert Trivers, who used the example of the weaning caribou calf to illustrate the logic of the conflict of interest between parents and children. In the

---

1 Sarah Blaffer Hrdy, *Mother Nature: Maternal Instincts and How They Shape the Human Species* (New York: Ballantine Books, 2000), 484.

early days of breastfeeding, the mother's and baby's interests are aligned. Not only does breast milk provide the infant with nutrients, water, and other ingredients critical for survival, but breastfeeding prevents the mother from resuming reproductive cycles. If she resumes too soon, before the baby is ready to manage on its own, both mother and baby suffer the costs—the baby may die, which for the mother means losing the huge investment that she has made thus far. That said, once the child can manage on its own or with someone else's help, it is in the mother's best interest to resume cycling and start on her next baby. In other words, she should wean the current child. But for a period of time, that child disagrees. It wants to stay in the warm embrace of its mother, where it both continues getting nutrition and delays the arrival of the competition—a sibling. You would expect this disagreement over weaning to be heated in humans because the weaned human baby still needs many more years of care and protection, and the next baby might have a different father—conflict is expected to be more intense between individuals who share fewer genes. Trivers argued that behaviors observed in mammalian babies, including human ones, such as tantrums and psychological manipulation at weaning time, evolved to further the child's own interests at the expense of the mother's.

The subject of weaning conflict calls to mind those cases of mothers who breastfeed their children past the age of five, ten, or even older. Is extended breastfeeding driven by an especially tenacious child or perhaps a mother who is more indulgent than most? Can breastfeeding a child turn into incest? There aren't clear answers to these questions, but that doesn't stop folks on Twitter from offering fierce opinions when a brave soul posts a photo of their ten-year-old suckling at their breast.

More generally, it can be reasoned that whining, crying, having tantrums, and engaging in other attention-getting strategies that many young children employ, not just at weaning time but *all* the time, evolved to divert maternal attention away from other distractions (e.g., siblings or the mother's own maintenance) back to the child in question. In my house, I'm often fielding escalating demands from multiple kids simultaneously. These behaviors extend to overt sibling rivalry, which, I know from my own experience and from talking to other moms, can dominate family life. Many studies on sibling rivalry have been done in birds that lay large clutches of eggs. The hatchlings beg for food with brightly colored mouths and loud vocalizations, and often the squeakiest bird gets the worm. In some species, this conflict extends to the practice of siblicide, in which stronger siblings peck their weaker ones to death, which is frequently the case in eagles, owls, and herons. The obvious loser in these examples is the sibling that doesn't get the worm or is killed, but the mother also loses. She would be twice as successful if both the bully and the bullied survived.

While these examples highlight rivalry, remember that siblings also have shared interests since they do share genes and in some contexts are expected to cooperate. One recent study was done on burying beetles, the parents of which provide varying levels of care to their larvae. In an elegant experiment that you can do in beetles but not mice or humans, two populations of beetles were allowed to evolve over many generations, one with parents providing care and the other without any care (the parents were removed in each generation before the eggs hatched into larvae). At the end of the experiment, the results were striking. In beetles with parental care, siblings evolved to be highly competitive—the strategy that worked best for beetle larvae was extracting more care from parents at the expense of siblings. However, when no care was available, siblings evolved to be cooperative. Without parents in the picture, baby beetles did better relying on each other. Maybe the solution to all the bickering between our kids is to leave the house and let them fend for themselves.

To recap what we've covered so far, theoretical work and animal studies illustrate the sort of tricks kids might play on their moms to get what they want, including staying in the womb as long as possible, being flashy, being cute, eliciting the production of feel-good hormones in mothers, refusing to wean, being the squeaky wheel, and getting rid of the competition. Some of these tricks are likely harmless—they don't compromise the mother's overall reproductive success—and they may even boost it. When those feel-good hormones are released in a mother when her baby tugs at her nipple, she is encouraged to do it again, which benefits child and mother. Some of these tricks, however, may not be in the mother's best reproductive interests and were likely driven by conflict. Let's look at some of the direct evidence supporting this idea.

Remember the remarkable phenomenon of genomic imprinting, a system of inheritance that causes genes to be active in children in a parent-specific way. During imprinting, marks are deposited on certain genes in eggs and sperm that influence whether that gene will be active or not in the future baby. Most of our genes don't "remember" which parent they came from, but on the order of two hundred genes have been shown to be imprinted in humans. Many of them are active in the placenta and the brain. As we discussed in chapter 6 on pregnancy, the imprinting marks deposited by dad push for bigger placentas and bigger fetuses, and those deposited by mom favor restraint. But what's the evidence for this? And what happens after birth?

A convenient way to figure out how something works is to break or disrupt it, something my inquisitive brother did frequently as a child, resulting in many broken toilets, electrical shocks, and destroyed electronics. A strategy used in genetics to figure out how a gene works is to break or disrupt it in lab organisms like flies or mice. In lab mice, if you disrupt regions of the genome that are imprinted, some fascinating things happen to babies. When the mother's signals are disrupted, leaving only the father's interests to be expressed, babies are born much heavier than average. When the father's signals are disrupted, leaving the mother's signals in charge, babies are born lighter. We can't do these experiments in humans, but sometimes a gene is disrupted on its own in an individual or family. By observing differences in the individuals who carry the disrupted gene, we can deduce what the intact version is doing. A variety of human disorders result from disrupted genomic imprinting, providing some of the best evidence for pre- and postnatal mother-child conflict.

Beckwith-Wiedemann, Prader-Willi, Silver-Russell, and Temple syndromes are distinct disorders caused by disrupted imprinting in various locations of the genome. The specific disruptions to these imprinted regions reveal what maternal and paternal imprints are normally doing in children. It's useful to come back to the tug-of-war or arms race analogy here. Conflict theory predicts that moves in one party will be countered by moves in the other in an escalating conflict. In kids with both functioning copies of imprinted regions, the effects of individual imprints are somewhat obscured by opposing tugs from the other party. Only when a tug on one side is removed is the evolutionary compromise or standoff between parents disrupted, allowing us to deduce the roles of imprinted genes.

Beckwith-Wiedemann syndrome is characterized by fetal overgrowth, prolonged and intense suckling behavior, large tongues, and accelerated growth in infancy. The genetic disruptions underlying Beckwith-Wiedemann syndrome result in the domination of paternal signals—the mother's balancing input is lost or swamped out in these children, revealing paternal tugs in the evolutionary tug-of-war. Many of the symptoms of Beckwith-Wiedemann align with predictions of the conflict theory that paternal genes are greedy, attempting to extract more resources from the mother. In contrast, the genetic disruptions of Prader-Willi, Silver-Russell, and Temple syndromes leave maternal signals to dominate, revealing their thriftier goals. Babies with Prader-Willi have poor appetite, poor suckling ability, sleepiness, and a weak cry. Infants with Silver-Russell syndrome are born two to three weeks early and underweight, show little interest in nursing, and have poor suckling ability, a symptom also shared by infants with Temple syndrome. Interestingly, though, in children with Prader-Willi, symptoms start to shift around weaning, which is the time that human fathers and extended family start feeding children and children practice feeding themselves. Between the ages of one and six, those with Prader-Willi develop an insatiable appetite, begin engaging in foraging and food-stealing behaviors, and often become obese.

As Haig has argued, the symptoms of imprinting disorders support the idea that genetic conflicts between family members extend well beyond birth, influencing transitions in a child's life such as the timing of weaning and puberty. Children with Prader-Willi and Temple syndromes often hit puberty early, which may be evidence that maternal genes favor reducing the burden of older children by promoting their earlier sexual maturation. These life transitions are certainly also influenced/constrained by other evolutionary forces (such as kin cooperation and natural selection in response to environmental pressures), but the role of conflict is well supported by the symptoms of imprinting disorders discussed earlier.

Children in our evolutionary past evolved traits that helped them secure what they needed from their mothers, sometimes at their mother's expense. These traits—some irresistible, some exasperating, some treacherous—linger in our children today. Our responses to our children, which I'll consider next, are similarly influenced by the age-old negotiations that have played out between the genes of family members.

Tricks up mom's sleeve

In any conflict that unfolds in real time, such as one I have frequently with my daughter on what to wear to preschool in the morning, both sides engage. My daughter says she wants to wear her summery mermaid dress, an inappropriate choice given that it's December and she frequently trips on the long tail at the back of the dress. I say no. She tries to sweet-talk me. I almost give in—her perkiness and passion for fashion are adorable—but I come to my senses and say no again. She gets increasingly agitated. I get increasingly agitated. Tantrums ensue on both sides, until one of us capitulates or a compromise is reached. No, you can't wear the mermaid dress to school, but after school, you can wear whatever you want.

In the evolutionary negotiations between family members, children have evolved tricks, but so have mothers. The evidence

I presented above on disorders of genomic imprinting illustrates that it's not only genomic marks deposited by fathers that have been shaping our children's behaviors; mothers have been exerting their will too, as revealed by the symptoms of Prader-Willi, Silver-Russell, and Temple syndromes. For many of the evolutionary tugs from fathers, it appears mothers tugged back, attempting to put the brakes on child greediness that interferes with enhancing their overall reproductive success.

While the cases of conflict described earlier have played out between genes of maternal and paternal origin *within* children, influencing a child's traits and behaviors, other cases manifest in a mother's traits and behaviors with her children. In birds, mothers and hatchlings benefit from some amount of begging, a behavior that lets mothers know that their babies need food. But if the intense begging of one hatchling interferes with the survival of a sibling, begging behavior has likely also been shaped by conflict. Consistent with predictions of conflict theory, studies across bird species have shown that the intensity of begging correlates with how closely related nest mates typically are—more aggressive begging is found in species that exhibit lower levels of nest relatedness. Nest mates might not be full siblings because the mother mated with more than one male or because a nest "parasite," such as the brown-headed cowbird, deposited its eggs in someone else's nest.

Begging in birds is urged along by the hormone

testosterone. In some species, such as the pied flycatcher, it's been shown that higher testosterone is correlated with more intense begging and higher survival rates. While making more testosterone is likely an evolved chick trick to garner more attention from mothers at the expense of siblings (and thus mothers too), mothers in some species may have responded by depositing more testosterone in the egg yolks of chicks that will need it the most. In many species, mothers incubate their eggs sequentially such that the chick of the first-laid egg hatches first. By the time the last chick hatches, its older siblings are larger and have a competitive advantage. But in canaries, great tits, and black-headed kittiwakes, mothers help the younger ones out by depositing more testosterone in later-laid eggs, helping them beg more intensely and level the playing field.

In humans, maternal behaviors that level the playing field between siblings are more difficult to study. Mothers are often suspected of having a "favorite," and I'll never forget my mother's answer when I asked who hers was. She said it was the child who needed her the most, a cryptic and unsatisfying answer to me at the time. Many mothers, though, do seem to favor the neediest of their children. Perhaps a compensatory strategy, as observed in canaries, is behind this instinct? An extreme version of this is exemplified in the novel *My Sister's Keeper*, in which a mother determined to save her older daughter's life forces her younger daughter to supply the blood cells and organs needed to treat the older girl's leukemia. While speculative, it is plausible that this type of maternal favoritism is influenced by ancient maternal instincts to redress the balance between siblings. Importantly, though, these instincts are flexible and contingent on circumstances as we discussed earlier, so in a different story, the mother from *My Sister's Keeper* might have favored the healthy, stronger daughter. We observe this flexibility in the black-headed kittiwake mothers mentioned earlier, who compensate later-laid eggs with more testosterone only under certain conditions. If the second chick is very likely to survive (because conditions are great, such as the mother's health and the amount of food available) or very likely to die (because conditions are terrible), the mother doesn't bother adjusting testosterone levels in the second egg. But under uncertain or intermediate conditions, kittiwake moms give their younger chicks a boost with more testosterone.

Not all maternal tricks must come at someone's expense. We've discussed the hormonal changes that occur in new mothers, in particular an elevation of prolactin and oxytocin. Prolactin is the hormone that promotes milk production and helps suppress ovulation during breast-feeding, which benefits both mother and child in the early part of a child's life. Oxytocin nudges mothers to continue engaging in behaviors that are critical to the mother and child's success—it's like a trick mothers play on themselves. Children evolved to tap into this system with their strategic suckling, clinging, cuddling, and hand-holding (an activity which, for me, always brings a moment of bliss). And by the way, oxytocin is also released in the brains of children engaging in these behaviors. The result is a coevolved, cooperative system that benefits children and mothers. At least until the next conflict begins.

Thinking more broadly about the evolution of motherhood in our species, perhaps one of the biggest tricks of human mothers was getting fathers and other family members involved in caring for our children. Mothers in most mammalian species shoulder all childcare responsibilities. I mentioned chimpanzee mothers who carry, protect, and breastfeed their babies for five years, after which time weanlings feed themselves. Chimpanzee fathers contribute almost nothing during this time (but see below), and the same is true of most primate fathers. In contrast, human fathers and extended family help feed and raise children. In other words, humans are cooperative breeders. I'll discuss the contributions of other family members in the last chapter, but here I'll focus on fathers.

The evolution of paternal care in humans and a related trait, pair-bonding, have been much debated in the human evolution literature. Paternal care describes any investments that fathers make in their children, whether in the form of food, protection, physical or emotional engagement, or financial support. Pair-bonding is a marriage of sorts between a female and male (not necessarily lifelong or exclusive), and it is often observed in species with paternal care (but not always). Both traits are rare in mammals. Because of long pregnancies and lactation in mammals, often the better reproductive strategy for males is to pursue additional mating opportunities rather than stick around to help the current mother and child(ren).

Pair-bonding and paternal care vary in men—some are involved partners and parents, while others are not. A fascinating body of work shows that the same hormones involved in maternal behaviors—prolactin and oxytocin—are also involved in paternal and pair-bonding behaviors in a range of species, including humans. These hormones are generally higher in males of species that have pair-bonding and paternal care (like prairie voles and humans) compared to those that don't have these traits at all. Also, *within* a species, such as our own, especially devoted dads and partners seem to have higher levels than uninvolved males. These observations have prompted clinical experiments in humans on the effect of nasal oxytocin sprays on male parenting (and other social behaviors), with some positive effects. Despite variation within humans, pair-bonding and paternal care are part of the human behavioral repertoire and were likely exhibited to some degree by our ancestors. These traits dramatically changed the lives of females, who would have been relieved of some of the heavy burdens of child-rearing while boosting their own reproductive success. How and why did these traits evolve in the human lineage?

The key to understanding the evolution of these traits

in our human ancestors may be looking at a trait that is *not* exhibited by human males—infanticide. As we've discussed, males in many primate species target and kill the breast-feeding infants in a group they are trying to take over, often with their sharp canines. While humans (women and men) do kill and abandon infants, the reasons humans do it are different and more varied than those of a male primate. Male primates have a very specific goal: to hasten females' return to menstrual cycling and ovulation, making them available for mating. Infanticide is prevalent in primates because lactation periods are especially long and because of the social structure in many species—groups are controlled by one or a small number of males who are continuously challenged by nondominant males who want to reproduce. Infanticide is an extreme manifestation of male-male competition that is devastating for female primates (and for species as a whole—it is not difficult to imagine how a population will fare when up to 50 percent of babies are killed by infanticide).

In species such as langur monkeys, Japanese macaques, and chimpanzees, one evolutionary response to infanticide taken by females is mating promiscuously to confuse paternity, as males are less likely to kill babies that might possibly be theirs. Concealing ovulation in some female primates (including langurs and humans), as opposed to advertising it with sexual swellings and bright colors, was an additional measure likely taken to confuse paternity. A more effective evolutionary solution to infanticide observed in some primates may have been pair-bonding and paternal care. In a comprehensive study of 230 primate species exhibiting some combination of infanticide, pair-bonding, and paternal care, a clear pattern emerged: Infanticide always comes first. In other words, in primates at least, pair-bonding and paternal care do not evolve unless the pressure of infanticide is there.

Ornithologist Richard Prum, the ardent supporter of Darwin's aesthetic theory of mate choice that I discussed in chapter 5, has argued that the infanticide problem in our human ancestors was solved through the mechanism of female mate choice. He speculates that our female ancestors (of over 4 million years ago) began choosing less aggressive mates with smaller canines, which over time reduced the incidence of infanticide and sexual coercion as well. He takes this idea further, hypothesizing that with their advanced sexual autonomy, our female ancestors began choosing males who made better partners and fathers, transforming the male sex, human families, and ultimately the human lineage. These are compelling, albeit untested, ideas.

Another possibility is that males and females began associating with each other to protect their nursing infants from infanticidal males. In fact, these associations are observed in our closest primate relatives, chimpanzees, who live in groups of multiple males and multiple females with no pair-bonding or much paternal care. A recent study showed that in contrast to previous reports, males do provide a minimal amount of care to their offspring in the form of protection. They associate more with their own infants and the mothers of those infants than they do unrelated offspring, but only when the infanticide risk is high, i.e., in the first one to two years of life. This minimal amount of paternal care—chimpanzee males don't do much else—might explain why infanticide isn't more rampant in this species. In the human lineage, mother-father associations that initially existed to protect infants from infanticidal males may have then developed into longer-term bonds and higher levels of care from fathers, ultimately eradicating infanticide—the trait does not exist in any primate with pair-bonding and high levels of paternal care. Precisely how and when this happened and what other factors were involved remain unclear.

For whatever reason pair-bonding and paternal care initially evolved and were subsequently maintained in humans, there is no question that these traits were game changers for the human species. More investment by fathers contributed to human birth intervals shortening, childhoods lengthening, and brains expanding, which allowed for the evolution of increased social, intellectual, and cultural complexity in humans. For our female ancestors, putting an end to infanticide and enlisting males in family life were the ultimate evolutionary tricks, contributing to their increased reproductive output. This new family arrangement had far-reaching consequences, contributing not only to female reproductive success but also to the success of the species as a whole.

Circling back to the topic of modern motherhood, how do the evolutionary tricks that we've been discussing help us interpret and navigate our own experiences as mothers and as the children of mothers? What does all this mean for the conflicted mother?

For myself, when my kids do something exasperating, like wake me up at 3:00 a.m. because of a nightmare, rat out a sibling, or whine incessantly about something trivial, I remind myself that many of these behaviors evolved in a different context, when they made the difference between life and death. The irrational fears our children have—of the dark, of monsters, or of being left at day care—are relics of a time when the dark was dangerous, there were scary creatures lurking, and being abandoned was more than just a remote possibility. The sibling rivalry that dominates my household exists in part because children in the past who secured a larger share of limited resources from their parents were the ones who lived long enough to become our ancestors. That said, the overlapping genetic interests of my kids should allow for more cooperative relationships, if I could just figure out how to draw those out.

Likewise, while I don't question those moments when I'm feeling especially maternal, I am often consumed with guilt when I prioritize my own needs and desires. In these cases, I remind myself that for hundreds of millions of years, mammalian mothers have been balancing their own bodily maintenance with the needs of their children. Successful mothers in the past were not unconditionally nurturing and selfless—they were flexible strategists, evaluating the ever-changing context and making decisions that were in their best interests. As Hrdy has argued, our feelings of guilt exist in part because of societal views on motherhood, which themselves have been influenced by the ongoing evolutionary tensions between females and males. Since the beginning of separate sexes, mothers and fathers have been in a tug-of-war over who should put in the childcare work and how much, and societal expectations that mothers be selfless reflect some of the recent tugs in this ongoing conflict. When I'm feeling guilty for saddling my husband with onerous childcare duties, I remind myself that human fathers are supposed to help—the human species wouldn't be where it is today if they didn't. And while I like to give my husband a hard time, in truth he's an excellent father, sharing the responsibility of raising our four kids. He's the one dealing with most of the nightmares in the middle of the night. So going back to my favorite Greek myth...you might say that in our species, mother and father have evolved to hold up the weight of the heavens *together*.

CHAPTER 8

In Sickness and in Health

A popular premise in literature and film is the apocalypse, whether brought about by aliens, war, or a deadly virus. When the coronavirus pandemic of 2020 hit, I found myself dusting off the covers of the apocalyptic novels I've read over the years, especially those based on a viral outbreak. A classic from the '70s is *The Stand* by Stephen King, which opens with the accidental release of a flu strain that was developed as a biological weapon. More recent favorites are *The Dog Stars* and *Station Eleven*, both of which are set primarily in the aftermath of global pandemics that wipe out most of humanity. As a lover of fiction, I'm drawn to the dark, poetic language in these novels and their philosophical reflections on what it means to be human. As a biologist, though, I can't help but return to the same geeky question as I read these books: What kind of biological factors might provide natural protection against a deadly virus, like the few survivors in these books are presumed to have?

The coronavirus strain that causes COVID-19 is nowhere

near as lethal as the pathogens in these works of fiction, but it has caused the deaths of millions and chronic illness in many people. As in the novels I mentioned, some of us in the real world have innate protection from COVID-19. Before vaccines were developed, it became clear that there were resistant individuals who never got sick or tested positive for the virus, even in the face of high exposure at work or at home caring for sick relatives. Also interesting are those who do become infected (as revealed by a positive test) but remain symptom-free. And there are those who become only mildly ill and bounce back quickly. Many protected adults are women. The geeky biologist in me is questioning why. Why are some people resistant to infection or protected from illness while others become severely ill? A related question pertains to the original host of the virus that causes COVID-19, bats, from which the virus jumped to another animal (possibly pangolins) and then to humans. Why does this virus make humans so sick but not bats, which appear to live with the virus without many health problems?

Scientists have been asking these kinds of questions for decades, not only about COVID-19 but about many other diseases, including malaria, tuberculosis, HIV, cancer, heart disease, and diabetes. For each of these diseases, some folks are more susceptible than others. Lifestyle and environmental factors play a role in these differences, but genetic and biological factors are also involved. The picture that has emerged is that our biological makeup is an important factor influencing if and how we get sick, a central tenet of the field of personalized medicine.

Each person's biological makeup has a long evolutionary history, spanning hundreds of millions of years ago to today. Some features of our biology differ by sex. Some features differ among individuals because of rare mutations that pop up by chance in our genomes. Some features differ among ethnic groups because each group shares a recent evolutionary history. How do these features influence human health and disease? How does our evolutionary past affect our health today?

In this chapter, I will tackle this question in reference to all humans, but in keeping with the theme of this book, I will focus on women. I group the chapter into four categories of diseases that afflict humans: infectious diseases, autoimmune diseases, degenerative diseases, and cancer. These categories are not mutually exclusive—for instance, infectious pathogens are known to increase risk for certain cancers and other degenerative diseases—and I am neglecting some, such as neuropsychiatric disorders (including addiction, depression, and autism). As you will see, not only do evolutionary insights help us understand how and why we suffer from these diseases, but they also give us promising leads on treatments.

Our perennial relationship with infectious pathogens

Humans have been living with infectious disease for as long as we have existed. The virus that causes COVID-19 is one of many types of viruses that infect humans, and viruses are one of many types of pathogens with which humans must contend. These pathogens, including viruses, bacteria, fungi, and parasitic worms, need a host for their life cycle, and the successful ones spread from host to host. Just as the virus that causes COVID-19 spreads from one human to another when we come in close physical contact with each other, all infectious pathogens have evolved strategies to spread from host to host. Intestinal parasites shed their eggs in feces, which, although gross to think about, can be a conduit into the gastrointestinal tract of a new host.

Our natural response to a pathogen is influenced by our own genetic makeup but also by the genetic makeup of the pathogen. Some pathogens, like the viruses that cause the common cold, have been infecting humans for tens of thousands of years. Both our genome and those of cold viruses have been coevolving for that time in a tug-of-war like others we've discussed in this book. While we'd all rather not get colds, we've come to a compromise of sorts with cold viruses, written in our genomes (both human and cold virus), that allows them to carry on their business without making us deathly ill in most cases. Severe sickness or death isn't good for the host (obviously) or the virus, which needs its host to be alive and mobile to help spread it to another person.

Other pathogens, however, like the one that causes COVID-19, are novel for humans. They jump from one species, in this case bats, to ours, bringing with them the compromises they had with their original host. Evolutionary compromises with bats are unique because bats are unusual mammals. They fly, they live in densely populated habitats, and they sing to communicate with other bats, all factors that help spread viruses between individuals. Consequently, bats carry more viruses than other mammals.

The adaptation of flight required some modifications to the bat immune system to handle all these viruses. Most mammals (including humans) use inflammation as a major defensive strategy against pathogens. If you've ever had a cut in your skin, you're familiar with inflammation—the area around the cut gets red and tender, because your body has sent inflammatory cells to the site to trap germs and heal the wound. This happens inside your body too when it's fighting infections. But the physiological requirements for flight include *low* inflammation—it's hard to fly when your tissues are inflamed—so bats focus on other strategies to control their viruses. They have amped up and expanded their use of interferons, proteins all vertebrates use to help control infection. Bat viruses, in turn, have evolved to suppress or subvert interferon activity in an ongoing evolutionary escalation. The upshot here is that the virus that causes COVID-19 and bats have coevolved to live somewhat peaceably with each other, the way we do with common cold viruses, but the virus provokes a huge and sometimes lethal response in humans because our immune systems have little experience with this type of pathogen (and the virus has little experience with us). Genetic studies of those who have gotten critically ill from

COVID-19 have shown that a few people have genetic errors in interferon genes, highlighting how important interferons are (whether in bat or human) in controlling this virus.

In contrast to individuals who have rare mutations that make them *more* susceptible to COVID-19, some individuals have natural resistance to the virus, like the characters in my favorite end-of-the-world novels. A team from Brazil has recruited couples in which one person has gotten severely ill and the other is resistant. They've compared the genomes of these couples, finding immune gene differences that may play a role in their responses to the virus. Because the virus that causes COVID-19 is a relative newcomer, we know less about the gene variants that provide resistance, but studies on a range of other infectious diseases have found specific gene variants that provide resistance to infection or at least protection from severe illness. As an example, a variant was discovered in a man in the '90s that confers resistance to HIV infection and another that provides some protection from severe hepatitis C illness. These kinds of natural variants are very rare, the result of random mutations in the few people in the world who harbor them. They're like the random mutations in the peppered moths from the introduction that make light-colored moths darker.

Also interesting are more common protective variants found in individuals with shared ancestry. To be clear, over 90 percent of all the genetic variants that exist in humans *don't* cluster by geography or "race"—they are found in many or all regions of the world, which shows that race is primarily a social, not a genetic, construct. But a small percentage of variants are associated with the geographical origins of our ancestors. The classic example is the variant in the hemoglobin gene that provides some protection against malaria but if inherited in two copies leads to sickle cell disease. At some point long ago, this variant was rare like that for HIV mentioned above, but it increased in frequency in individuals from Africa and Southeast Asia, where malaria kills millions of people each year. It is absent from populations living in regions of the world without malaria. The variant is beneficial and persists only where malaria is present, even with the higher risk of developing sickle cell disease. Another example are ABO blood types, which are important to match correctly during a blood transfusion, but they also influence susceptibility to infectious diseases like meningitis, cholera, and plague. Blood type frequencies vary around the world, suggesting that specific blood types are common in places where they helped (and may still help) defend against region-specific pathogens. Many studies have suggested that blood type O offers slight protection against COVID-19 illness.

East Asians have a suite of immune gene variants that likely protected against *ancient* coronavirus infections. These variants, which are missing from other populations in the world, suggest an evolutionary arms race with coronaviruses that started at least twenty thousand years ago in Asia. Although it's not clear if the variants provide natural protection against COVID-19 illness, the discovery points to several new genes that might provide novel drug targets. Going even further back in time, in research that sounds more like science fiction than science, we know that modern humans and Neanderthals had sex in places they overlapped in Europe and Asia (on the order of fifty thousand to one hundred thousand years ago), which explains why 2 percent to 3 percent of modern European and Asian DNA is derived from Neanderthals. Humans and Neanderthals didn't just exchange DNA; they also exchanged viruses. It turns out that many of the Neanderthal DNA sequences that stuck around in Europeans and Asians are immune gene variants, which likely protected Neanderthals against viruses they had been living with for thousands of years but were novel to modern humans. This has been described as the poison-antidote model: When humans interacted with Neanderthals, they were exposed to a poison (Neanderthal viruses), and the Neanderthal DNA they retained in their genomes (Neanderthal immune gene variants) was the antidote.

Although genetic variants, rare and common, may influence our response to infections, many other factors are involved. A solitary person who rarely leaves their house is less likely to be infected with a pathogen. Younger individuals are less likely to become severely ill from most pathogens than older ones, something I'll return to later. Being healthy overall improves one's chance of bouncing back quickly after an infection. And it turns out that women are more protected than men from severe illness during infections. In the case of COVID-19, men compose more of the serious cases and deaths than women, even though women and men are vaccinated and become infected at roughly the same rates. For every ten women who die of COVID-19, fourteen men do. Let's finish this section by discussing why.

Recall from chapter 1 the distinction between biological sex and gender, both of which have been shown to affect disease outcomes. Gender influences our risk of exposures, our access to health care, our use of the health-care system, and our treatment within the health-care system. Biological sex influences our hormonal and immune profiles, which affect how our bodies respond to infections. In general, females have faster and stronger immune reactions than males, which are mediated in part by sex hormones. Estrogen heightens the immune response, and testosterone dampens it.

Why would males want to dampen their immunity? The short answer is that you can't have it all. Males in most mammalian species invest much energy into building muscle and in costly behaviors like showy displays and aggression toward other males. In my pubescent son's case, these behaviors include extreme mountain biking and bike jumps that make my heart leap out of my chest. These energetically costly activities must come at the expense of something else, in this case immune defense, which is another costly endeavor. For females, who have evolved to invest more in bodily maintenance to support long pregnancies and lactation, heightened immune responses help fight infection better. Studies on COVID-19 specifically have shown that one sex difference that might play a role in disparate outcomes is a protein that sits on the surface of cells and allows the virus inside. The activity of this protein seems to be dampened by estrogen, whereas men have higher activity and levels of the protein, which may increase their chance of severe sickness and death.

To recap, humans have a long evolutionary history with infectious pathogens. Biological and nonbiological factors influence how we respond to them. Biological sex is one factor, with the immune system operating slightly differently in males and females because of evolved differences in reproductive strategies. Rare genetic mutations, which are an important source of natural variation in any population, can influence our susceptibility to and protection from certain pathogens. Ancestry is another factor. As modern humans populated the globe, they faced different pathogens, and their genomes evolved to deal with those pathogens accordingly. The way humans dealt with pathogens in the past left lasting marks on our genomes. These lasting marks influence human health and disease today, which isn't always a good thing, as I'll consider in the next section.

When the enemy is your own immune system

For women who are feeling superior about their robust immune systems, the truth is they are a double-edged sword. The female immune system tends to work better when fighting infection, but sometimes it becomes hyperactive, which can lead to autoimmune diseases, characterized by the body's immune system attacking itself by mistake. These diseases are not major causes of death but are painful and debilitating, including lupus, Graves' disease, rheumatoid arthritis, asthma, Crohn's and other inflammatory bowel diseases, and multiple sclerosis (MS). About 8 percent of humans suffer from auto-immune disease, and 80 percent of all cases are found in women.

Unlike infectious diseases, which humans have always lived with, autoimmune diseases would have been extremely rare in our hunter-gatherer ancestors. They are diseases of the modern world. What has changed? In the last few hundred years, humans have experienced rapid cultural and environmental changes, including largely beneficial changes like antibiotics, vaccines, and more hygienic living conditions and harmful ones such as richer diets, more sedentary lifestyles, and exposure to toxic chemicals. Before the Industrial Revolution, most people died from infectious diseases and at younger ages—infant mortality was especially high. Now, with modern medicine and hygiene, most infants survive, and more people live longer. But there are some unintended consequences of our newly sterile world. Autoimmune disorders may have increased in prevalence because of the cleanup, an idea referred to as the "hygiene/old friends hypothesis." What's the connection between our cleaner bodies and immune systems gone haywire?

In the past, our ancestors were chronically infected with bacteria, viruses, and parasitic worms. Some of these may have killed their human hosts quickly, but many pathogens take the long view, evolving strategies to persist longer in their host to produce more offspring. Much evidence points to parasitic worms as the missing suspect responsible for the increased prevalence of autoimmune disorders today. These parasites include hookworms, whipworms, and tapeworms, pathogens we in the United States don't worry about until we're traveling to a country in which they are still prevalent. There is an intriguing correlation between parasitic worms and immune disorders. In parts of the world with modern hygiene and sanitation, autoimmune diseases have skyrocketed. Many tropical countries, on the other hand, which continue to have widespread worm infections, lack autoimmune disorders.

Beyond these correlations, we now understand some of the biology of worm infections that explains why removing them from our bodies may have led to a higher incidence of autoimmune problems. Parasitic worms produce molecules that suppress immune and inflammatory responses in the host to avoid being killed. Human hosts who were chronically infected with worms also benefited from dampening their own immune response against worms to avoid suffering the damage of chronic inflammation. As with other relationships we've discussed in this book, the human immune system and worms have been engaged in a coevolutionary dance of sorts over immense timescales. However, when worms were suddenly removed, the equilibrium was disrupted, leaving humans with an immune system missing its long-established partner. Unfortunately, without this partner, the immune system can behave pathologically by targeting its own body tissues, whether it's the sheath around nerve cells in the brain and spinal cord (MS), cells in the gut (inflammatory bowel diseases), or cells in our airways (asthma and hay fever allergies). As mentioned, women have heightened immune responses in general, which explains why removing immuno-suppressive worms from our body ecosystems would result in a higher prevalence of autoimmune disorders in women.

Some observational studies on worms and autoim-mune symptoms support this hypothesis. In Argentina, MS patients who happened to develop worm infections went into remission for their MS, but MS relapsed when the worm infections were treated with antiworm drugs. Similar patterns have been observed with inflammatory bowel diseases and allergies. These observations led to trials in animals and humans on the effect of "worm therapy" on autoimmune symptoms, in which patients are infected with parasitic worms or eggs. Trial results have been mixed. In some, patients experienced improved autoimmune symptoms, or at least no worsening of symptoms, with worm therapy. Others, however, have reported placebo effects. Studies are now shifting to find molecules produced by worms that may be used as treatments for immune disor-ders, a more targeted approach that also avoids the practi-cal and ethical issues of live worm therapy. As someone who suffers from a variety of autoimmune issues, it's hard for me to imagine getting over the ick factor of being infected with live parasites. Depending on the worm, this is done by consuming worm eggs in liquid format or being injected with worms under the skin. A worm infection can be pretty nasty depending on the worm, and the cure for a worm infection if it gets out of control is also nasty. This means that treatment often consists of trying to kill the worms just before you kill the patient. No, thank you! I'd much rather take a worm-inspired drug that comes in pill format.

The takeaway here is that our ancestors were chronically infected with parasitic worms, and our genomes (human and worm) evolved a sort of compromise to live together. When we removed worms from our bodies in some parts of the world, we reneged on the compromise, but our genomes still haven't gotten the memo. Cultural changes, like better hygiene and medicine, proceed much faster than genomic ones. One unintended consequence of our newly sterile world is an increase in autoimmune disorders, characterized by the body's immune system attacking itself inappropriately. These diseases disproportionately affect women, who have more robust immune systems to begin with.

Diseases of the aging body

Recent cultural changes have had other consequences on human health, including an increase in chronic and degenerative diseases, such as hypertension, heart disease, diabetes,

Alzheimer's, and cancer. There are a couple of important issues here. First, better hygiene and medicine allow humans to live longer, so we suffer less from the infectious diseases discussed earlier but more from diseases that come with an aging body. Second, because of cultural changes related to diet and lifestyle (eating more and moving less), we suffer even more from chronic and degenerative diseases than our ancestors would have if they were lucky enough to reach older ages. But first, let's talk about aging.

Immortality is another favorite topic in literature, as in the classics *The Picture of Dorian Gray* and *Tuck Everlasting*. Sorry to burst any bubbles here, but evolutionary theory clearly explains why organisms in the real world must grow old and die. There are two parts to the answer. First, there are usually fewer older individuals than younger ones in a population because of death from infection, accidents, and other external causes. Without getting into the weeds, natural selection works better with bigger numbers, so it is less effective with increasing age, resulting in the age-related decline of the physiological systems that keep us going. Second, for genes that have varied jobs throughout our lives, a genetic change that boosts reproductive success early in life will evolve even if it accelerates decline later in life. A good example is the *BRCA1* and *BRCA2* mutations that increase one's risk of breast and ovarian cancer, made famous by actress Angelina Jolie after she had a double mastectomy in 2013. One study showed that among women who don't use birth control, those who have the *BRCA* mutations have more kids than those who don't have the mutations. This suggests that early life fertility benefits are the reason the mutations have stuck around in humans, even though they increase the chance of getting cancer later in life.

The main message is that aging doesn't have an evolved purpose. It is an inevitable consequence of other aspects of our biology. This doesn't mean we can't have healthier bodies and minds as we age—we can, which is the goal of a growing scientific community that studies aging and age-related disease.

One noteworthy fact about aging is that it differs between the sexes of most mammalian species, reflected in the average life expectancy of men and women in the United States: In 2021, it was about seventy-six for males and eighty-one for females. This difference in life expectancy can be explained by a few factors. First, as I mentioned, testosterone promotes costly behaviors in males that presumably evolved to attract mates or fight off other males. These testosterone-driven behaviors cause many more deaths of males than females under twenty-five. I try not to think about this when my son goes out on his reckless bike expeditions. But even when you account for more male deaths at these younger ages, the sex difference in life expectancy still holds. Women do live longer. As discussed earlier, males have compromised immunity compared to women, likely contributing to shorter life spans. Their cells and bodies also seem to age faster than those of women, for reasons that are being actively explored. Many of these differences appear to be linked to sex hormones. Estrogen is protective, and testosterone is damaging.

Paradoxically, a good illustration of estrogen's protective effects is what happens when women stop producing estrogen in large quantities. Before menopause, women are well-protected from heart disease and type 2 diabetes, two chronic diseases that are leading causes of death in humans. People with heart disease have a buildup of plaque in the arteries that supply the heart with blood, and those with type 2 diabetes cannot properly control the movement of sugar (glucose) between the blood and body tissues, leading to high blood sugar that ultimately damages blood vessels, nerves, and organs. Among its many functions, estrogen is involved in keeping arteries flexible and in controlling cholesterol levels; it also helps insulin regulate blood sugar more effectively. When estrogen wanes at menopause, a woman's risk of developing these diseases increases, especially heart disease. One in five women die from heart disease (compared to one in thirty for breast cancer, a disease that many women worry more about). These are staggering statistics that are underappreciated by women.

Alzheimer's is another disease against which estrogen provides some protection until menopause. Alzheimer's is one of the only major causes of death that hits women at higher rates than men. Age explains much of this difference. Alzheimer's develops in older individuals—anyone who survives to age eighty-five has a one in three chance of getting Alzheimer's—and women live longer. But Alzheimer's risk is still greater in women even after accounting for their longer life spans, likely because of the hormonal and brain changes that occur around menopause. Studies in animals and women have shown that the brain uses glucose less efficiently when estrogen levels wane, contributing to the increased risk of Alzheimer's in women. Another hormonal change at menopause—a large increase of follicle-stimulating hormone (FSH)—has recently been linked to the production of proteins in the brain that are implicated in Alzheimer's disease. In some exciting experiments on mouse models of Alzheimer's, blocking FSH reduced Alzheimer's symptoms (and other menopause-associated symptoms), opening up a potential treatment approach in humans.

To reduce all aspects of aging and age-related disease to our hormones would be misleading. As with the infectious diseases we discussed, many factors, biological and otherwise, influence how we age and fare against chronic and degenerative diseases. Another biological factor that may contribute to aging differences between males and females is their number of X chromosomes—males have just one, but women have two. How X chromosome dosage affects health and aging is being studied in some clever experiments on mice that allow researchers to study the effect of sex chromosomes independent of sex hormones. A biological factor that isn't sex specific is rare genetic variants. We discussed how they influence our immune responses to infectious disease, but

we eat and how much we exercise, which in turn influences our susceptibility to chronic diseases. Until recently, men in many parts of the world smoked more and ate greater quantities of red meat, contributing to a higher risk of heart disease at younger ages. These lifestyle trends have been converging in recent decades; for instance, fewer men but more women now smoke. In general, though, *both* men and women today have lifestyles that are vastly different from those of our ancestors, influencing our chances of developing the chronic diseases we've been discussing. How many times have you told yourself to eat healthier and exercise more? Our hunter-gatherer ancestors ate and moved very differently than we do today, and our bodies are adapted to this ancient lifestyle. In a short amount of time, we changed what we eat and how much we move, but genomes evolve much more slowly than culture. As a result, we suffer from obesity, hypertension, type 2 diabetes, and heart disease at higher rates, earlier ages, and greater severity. Let's discuss this in a little more detail.

Humans are somewhat unique in the animal world for being able to survive and thrive on a wide variety of foods, a flexibility that enabled humans to colonize the world. Some groups (in South Asia) are completely vegetarian, whereas others (in the Arctic) live mainly on meat. Nevertheless, ancient human lifestyles were characterized by high levels of physical activity and variation in food availability, whether that food was plant or animal or a combination. Sometimes food was abundant and sometimes it wasn't. Humans today move less and eat more, and for most, food is always readily available. The quality of our diets is also different, including highly processed grains, simple sugars, and foods high in saturated fats. Our bodies, though, are adapted to ancient diets and lifestyles, unable to keep pace with the rapid cultural changes of late. The imbalance between how much energy we expend and consume is one reason for the dramatic rise of obesity in industrialized countries and more recently in developing countries. In 2021, roughly 40 percent of Americans over twenty were classified as obese and 70 percent as overweight or obese. The wide availability of foods high in saturated fats and simple sugars also plays a role in the rise of metabolic problems. Heavy consumption of simple sugars may eventually lead to insulin resistance and type 2 diabetes, and a diet high in saturated fats and low in unsaturated fats contributes to heart disease. A related topic but beyond the scope here is our taste for and, in some cases, addiction to certain foods (and activities). The brains of our ancestors evolved to find that rarely encountered sweet berry, high-fat meat, or medicinal plant pleasurable and rewarding—those who consumed these things had better survival and more kids than those who didn't. But today this trait is a liability because these foods and substances are so highly concentrated and widely available.

To sum up, we live longer than our ancestors and our lifestyles are drastically different, which has led to an increase in both the incidence and severity of chronic and degenerative diseases. Women live longer than men, which reflects ancient differences in reproductive strategies between females and males. Women experience much pain and suffering as they age, especially when estrogen levels wane at menopause, a topic we'll return to in the next chapter.

I discuss one major category of age-related disease— cancer—on its own in the next section.

Cancer (and a surprising connection with the placenta)

Most people's lives have been touched by cancer, whether you've battled with it yourself or have a friend or family member who has. My father was diagnosed with lymphoma in his fifties and struggled with it for over twenty years, which opened my eyes to what a devastating disease it can be. After heart disease, cancer is the leading cause of death in both women and men. In the United States, over one-third of us will develop cancer in our lifetime, and one-third of those who get cancer will die from it.

Our chances of developing cancer increase as we age. Some of us are born with rare genetic variants that increase or decrease our susceptibility to specific cancers, as we've seen with other diseases that afflict humans. But it takes between five and ten mutations for a normal cell to turn into a cancer cell, and more cell divisions with age provide more opportunities for those mutations to accumulate. Also, many cancerous cells are destroyed by our immune systems, and our immune systems weaken with age. Like the degenerative diseases that we've already considered, cancer was certainly less common in our ancestors. They didn't live as long, and they weren't exposed to many of the agents known to increase one's risk of cancer today, like high-fat diets, tobacco, and environmental pollutants. As discussed in chapter 3, the use of birth control in women today is another risk factor for cancer. Our adult ancestors were usually pregnant or lactating, which means they had fewer menstrual cycles, fewer cell divisions in the breast, and thus fewer breast cancers.

For cancers that affect both sexes (that is, the nonreproductive cancers), men are more susceptible to developing and dying from cancer than women. Some of this may be due to gender constructs such as diet and behavior—men tend to have unhealthier diets and engage in riskier behaviors than women. Nevertheless, biology also plays a role. Because men tend to be larger, with more cells in their body than women, they have more cell divisions and thus are at greater risk of developing cancer. In the most common type of brain cancer, it was shown that male brain cells are more likely to transform into malignant cells than female brain cells. In liver cancer, it was shown that liver cancer cells are stimulated by testosterone and inhibited by estrogen. More generally, stronger immune systems in women may find and target cancers more effectively than in men, and the innate ability of a woman's cells (in the uterus) to restrain an invading fetus during pregnancy may also influence her ability to restrain invading cancer cells elsewhere in the body. This last idea is provocative and complicated, so let's break it down.

The big question we've been tackling in this chapter is how our evolutionary past influences human health and disease today. This past includes recent human evolutionary history, such as adaptations in our hunter-gatherer ancestors that protected against infectious diseases like malaria. But more ancient evolutionary history is also relevant, such as differences in female and male reproductive strategies that have existed since the beginning of separate sexes in our animal ancestors. Humans have an ape, primate, mammal, and amniote legacy, and traits we picked up from our long-ago ancestors may influence our health today. What does this legacy have to do with cancer?

For over a century, biologists have noted a similarity between the behavior of cancer cells and trophoblast cells of the placenta. Recall from chapter 6 that the placenta is made by the developing baby, with the job of extracting nutrients from the mother, exchanging gases, and removing waste. Trophoblast cells have many functions in the placenta, one of which is to invade the uterus during early pregnancy, moving through uterine tissue to gain access to and modify the mother's blood vessels. Cancers, on the other hand, are clusters of abnormal cells that form a lump or mass anywhere in the body, which become a concern if the abnormal cells start to grow uncontrollably and leave the mass to form new ones elsewhere, referred to as metastasis. It is the uncontrolled growth and spread of cancer cells that is so dangerous, because they interfere with the proper functioning of vital organs. So how are placental cells and metastatic cancer cells similar? They are both experts at invading other tissues, attracting the growth of nourishing blood vessels, and avoiding the immune system in the process.

Another observation about the placenta and cancer is that some species of mammals are more likely to develop metastatic cancer than others, and these mammals all have invasive placentas. Remember from our discussion of pregnancy that the mammalian placenta looks dramatically different across species. One of the ways biologists classify the placenta is based on how aggressively trophoblast cells travel into the uterus. Humans, bats, and rodents have deeply invasive placentas, whereas cows, pigs, and whales have superficial, noninvasive placentas. The species with superficial placentas suffer from fewer metastatic cancers than those with invasive ones.

Initially, the hypothesis was that metastatic cancer evolved when invasive placentas did. An unfortunate consequence of the evolution of invasive placentas was that cells anywhere in the body could now deploy the invasion program that was wired up during the evolution of pregnancy. However, this hypothesis is not supported by the data, which show that metastatic cancer long predates invasive placentas. Birds get metastatic cancers, but no bird species (extinct or extant) has ever made a placenta, let alone an invasive one. Consistent with this, it also appears that the mechanisms used by cancer and placental cells to invade are similar to those used in other very ancient processes like wound healing, which predate the origin of placentas by hundreds of millions of years.

There is another explanation for why mammals with noninvasive placentas have lower metastatic cancer rates. Recall from chapter 6 that when the mammalian placenta first evolved in the ancestor of marsupial and placental mammals, it was short-lived and relatively superficial. It became invasive in the ancestor of placental mammals, but it didn't stop changing then. In some groups, like the apes, the placenta evolved to be even more aggressive (which may explain why humans are so susceptible to metastatic tumors). But in the hoofed mammals, like cows, horses, and pigs, it *reverted* to being noninvasive. This isn't because trophoblast cells lost the ability to invade. For example, pig pregnancies that take hold at sites outside the uterus *are* invasive. What's different in these species is how the uterus responds to the invasion: It blocks the placenta from breaching maternal tissues. The maternal endometrium evolved resistance to invasion, possibly in the genetic conflict between mother and fetus over resource transfer, as discussed in previous chapters. Since endometrial cells are fibroblasts, a type of cell found all over the body, it is possible that the endometrium's evolved resistance to placental invasion in hoofed animals allows other fibroblast cells in the body to resist invasion by metastatic cancer cells, explaining lower malignancy rates in these species.

Some creative experiments in the lab support this hypothesis. A team led by my PhD advisor, Günter Wagner, added trophoblast cells from two species—human or cow—to a lawn of endometrial cells from human or cow, in different combinations. First, they found that cow endometrial cells resist invasion from cow trophoblast cells *and* human trophoblast cells, while human endometrial cells are more easily invaded by human *and* cow trophoblast cells. These results show that much of the control of invasion lies in the endometrium and that cow endometrial cells resist invasion better than human ones. The team went further and did the same experiments with cancer cells. They added human or cow skin cancer cells to lawns of skin fibroblast cells from human or cow, with similar results: Cow skin fibroblasts are more resistant to cancer invasion than human fibroblasts. They even identified some of the genes involved in this difference between species. When they tweaked those genes in human cells to make them more like cow cells, the human cells became more resistant to cancer invasion. This cutting-edge work suggests that a more detailed understanding of the genetic changes underlying placental evolution in mammals, especially how the endometrium has evolved, could lead to novel interventions to prevent cancer metastasis.

Now getting back to the provocative idea posed earlier. Men die from cancer at higher rates than women, and one of many possible explanations for this sex difference has to do with the endometrium, a tissue only females possess. Even though human endometrial cells aren't as effective as cow endometrial cells in restraining placental invasion, the human endometrium does provide some restraint—if it didn't, a woman wouldn't survive through a pregnancy. As discussed, endometrial cells are a special type of fibroblast cell that is found everywhere in the body. So there may be sex differences in how fibroblast cells in men and women oppose metastatic invasion, the roots of which lie in the female endometrium. These ideas have not yet been properly explored, but in light of centuries of male-dominated medical research, it is ironic that the endometrium—a female tissue—holds so much promise in improving cancer outcomes for all humans.

———

I'd like to share a conversation I had in my twenties with my dad about the scientific research I wanted to pursue. I told him I was joining a PhD program in evolutionary biology instead of going to medical school or doing biomedical research. While he was very supportive of me and my ambition, he did not understand my choice. Our conversation went a little like this:

"Will you make any money doing that?" he asked.

"No, not much."

"Will you cure cancer doing that?"

"No, probably not."

When I told him a couple of years later about my doctoral thesis on the evolution of the placenta, he pressed me again on why I was pursuing a topic without biomedical relevance.

At the time, my answer was simply that I thought the topic was fascinating. But over the years, I've come up with better answers for the skeptics.

Hopefully you're now convinced that our evolutionary past does impact health and disease today and that evolutionary research is an important complement to traditional medical science. I wish I could tell my late father that work related to my doctoral thesis on endometrial evolution is opening new avenues of research on treating metastatic cancer. Or that investigations of the coevolution of parasitic worms and humans is leading to more effective treatments for autoimmune disorders. Or that studies of old battles between humans and ancient coronaviruses may lead to better treatments for COVID-19. I'd even dare to say that understanding the coevolution of viruses with animal hosts like bats, which have been the source of many deadly viruses in human history, may prevent future outbreaks like those in the apocalyptic novels on my bookshelf.

CHAPTER 9

Menopause and the Epilogue of Life

Like many kids who grew up in the '80s and '90s, I watched a huge amount of TV, much of which was age-inappropriate. I don't know why *The Golden Girls* appealed to my preteen self, but I watched every episode. The one called "End of the Curse" was certainly over my head—at the time, I didn't know what the "curse" was, let alone the end of it. In the episode, Blanche thinks she's pregnant and shares her worries with the girls, who console her and tell her they'll help with the baby. But then Blanche goes to the doctor, who breaks the news: It's not pregnancy, it's menopause. This sends Blanche into a tailspin about losing her sex appeal and aging like her mother. Fortunately for Blanche, in a crazy subplot that involves breeding minks for fur coats, a cute vet comes to the house to examine the uncooperative minks. The vet tells Blanche that while the minks are too old to breed, Blanche has still got it. He asks her out on a date and all is well.

Just as the symptoms of menopause confuse Blanche

during this *Golden Girls* episode, the trait of menopause is puzzling on evolutionary grounds. We know that the winners of the evolutionary game of life are those who leave more gene copies in the next generation. So why would females in an entire species stop playing partway through the game? Men don't need to stop playing—they continue making sperm until the end of their lives. Females in most mammalian species don't stop playing either (we'll talk about those minks later in the chapter). What's going on in human women? To use a different metaphor, if human female reproductive life is the story, why is the epilogue so long?

In this chapter on menopause and longevity, I'll explain the simple reason we experience the end of the curse: We run out of eggs. I'll discuss the consequences of egg depletion on the rest of our bodies—what happens to us hormonally and physiologically during the menopause transition and beyond. But our main focus will be on the ultimate, evolutionary reason we run out of eggs when females in most other species don't. Some argue that a long life after menopause evolved by kin selection—postmenopausal grandmothers boosted their reproductive success by helping care for their grandchildren. Some claim that menopause evolved because of conflict, rather than cooperation, between older and younger females. Some claim that menopause is just a consequence of humans living longer. Some believe that it was a combination of these mechanisms. As you'll see, the debate hinges on two big unknowns: how long our ancestors lived and the structure of their families, questions that are difficult to answer for humans who lived so long ago.

What you're born with is what you've got

In previous chapters, we discussed when women produce eggs—or more precisely, when they produce ovarian follicles, each of which consists of an egg and supporting cells. In contrast to men, who continuously make sperm throughout their lives, women make all their follicles while developing in the womb. It's kind of mind-blowing to trace the history of the follicle that made you—it was produced in your mother when your grandmother was pregnant with her. This thought always makes me feel more connected to my grandmother.

Once produced in the fetal ovaries, some follicles—called primordial follicles at early stages of their development—stay dormant until they are needed over a decade later, at which point they continue their development and are ovulated. However, over the course of your life, starting when you are still a fetus, a fraction of your primordial follicles are continuously being awakened. Before you hit puberty, these awakened follicles are destined to die in a process called atresia, since follicles need hormone signals to complete their development. (After you hit puberty, some of these awakened follicles get recruited for further development during your menstrual cycle—I'll discuss this later.) The extent of follicle death by atresia is also mind-blowing.

We have a maximum of about 6 to 7 million follicles at twenty weeks of fetal life; when we are born, that number has been reduced to roughly 1 million; at puberty, we are left with about 400,000; and at menopause, we've got few to none left. We lose over 80 percent of our eggs before we've even hit puberty (!) and ovulate fewer than 500 eggs between puberty and menopause, so the *vast* majority are lost by atresia.

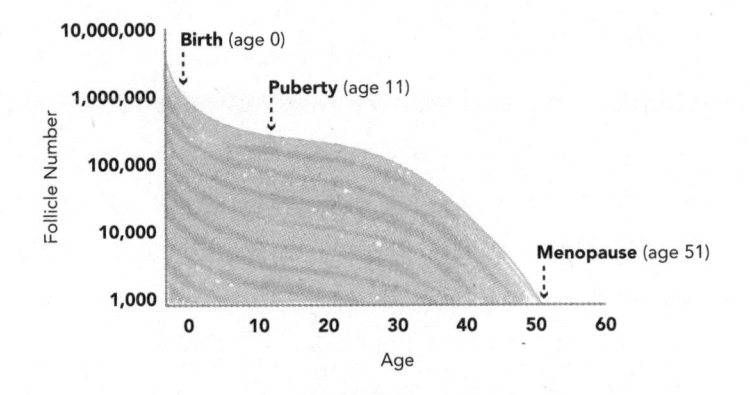

The loss of eggs throughout human life.

Illustration based on original by Healthily

To understand human menopause, the first question to answer is why we use this stockpiling strategy in the first place rather than the male strategy of making sperm throughout life. Any woman in her late thirties or forties who wants to have a baby has thought about the unfairness of this dichotomy. It turns out that making a fetal egg stockpile is not unique to human women—all female mammals and birds do it. On the other hand, females of other vertebrate species, such as fish, amphibians, and reptiles, continuously make eggs throughout their lives, similar to males of *all* vertebrate species, who continuously make sperm. Why are there two strategies, and why don't we use the one that would allow us to easily have children in our forties and fifties?

As I mentioned in previous chapters, female mammals (and birds) invest a huge amount of time and energy in reproduction. A female mammal has a lengthy pregnancy and then, in many cases, nurses her offspring for even longer; a female bird incubates her eggs and then cares for young hatchlings. Consistent with this large investment, the female has evolved quality control strategies that allow her to direct investment to the highest-quality offspring, those who will maximize her reproductive success. The uterus screens out poor-quality embryos during early stages of pregnancy. A mentor of mine at Yale, Stephen Stearns, first suggested that the dramatic culling of eggs—atresia—is an earlier form of quality control, selecting only the highest quality of eggs for ovulation; recent lab experiments lend support to the hypothesis and identify some of the mechanisms involved. Recent work has also shown that the female reproductive tract selects for better sperm, acting as a gauntlet through which sperm must pass to get to the egg. And relevant to the question at hand—why do we make our lifetime supply of eggs as a fetus?—the strategy of stockpiling is likely also for quality control.

The reason men are able to continue making sperm into old age is that they've got sperm stem cells that continue to divide over their lifetimes. The downside to the continuous production strategy is that every time a cell divides, there is an opportunity for a DNA mutation to arise. Mutations are often harmful, and we know that most mutations enter the human gene pool through sperm rather than eggs. For a typical male mammal that contributes nothing to childcare, the costs of these mutations (potentially lower-quality offspring) don't outweigh the benefits of continuous sperm production (more offspring). But for a female mammal or bird whose reproductive strategy is to heavily invest in fewer, higher-quality offspring, producing a stockpile makes more sense. There are over ten thousand mammalian and avian species on the planet, and not one of them has reverted back to the original strategy of continuous egg production.[1]

Not only do female birds and mammals make a stockpile, like humans, but they also deplete their stockpile in the process of atresia and to a lesser extent during ovulation. The consequence of having a stockpile that is depleted by atresia and ovulation is that if you live long enough, you will run out of eggs. But this is not an insurmountable evolutionary problem.

---

[1]   There are two possible reasons for this. One is that stockpiling continues to be the better strategy. The other is that the developmental process of making eggs is too difficult to tweak—the shift from a stockpile to continuous production would involve infertile or less fertile females. Either (or both) of these explanations is plausible.

Natural selection can act on the size of the stockpile—making more eggs up front—so that reproductive span ends around the same time as total life span. It can also act on the rate of egg depletion, perhaps by waking up fewer eggs each month or by tweaking the quality control process. Studies of stockpile sizes across mammals have shown that most mammals make a stockpile large enough to last their entire lives. Long-lived elephants, who reproduce into their sixties, have enough eggs to last their entire life. Fin whales live and reproduce into their nineties! Another interesting species is the naked mole rat, a blind, hairless, subterranean African rodent that lives decades longer than other rodents. Naked mole rats are eusocial, like honeybees, and the queen stays fertile into her thirties. A recent study showed that the naked mole rat produces an egg stockpile an order of magnitude larger than that of other similarly sized rodents, which supports the idea that natural selection has acted on the size of the stockpile in this species to ensure that the female has enough eggs to last her long life.

Humans stand out because women don't have enough eggs. It would appear that natural selection has not operated in our species to increase the egg supply up front or to reduce the rate at which we deplete our supply. But humans are not the only ones. In contrast to the long-lived and long-reproducing baleen whales, a few species of toothed whales are notable exceptions (and there may be more that have yet to be discovered). Killer whales, beluga whales, narwhals, and short-finned pilot whales experience menopause and then live for decades longer. Killer whales experience menopause in their forties but live into their eighties or nineties. Human women, then, are not alone in our experience of menopause, which informs our conversation about the evolutionary significance of menopause, as we will get to later.

The relationship between menopause, longevity, and aging

Our closest living relatives—chimpanzees—undergo menopause at about fifty, but in the wild, they rarely live into their forties or fifties. This has suggested to some anthropologists that menopause at fifty is not an evolved trait in human women; rather, our longevity is. In other words, we inherited our reproductive span from our ape ancestors, and what changed in humans is how long we live beyond menopause. Whether or not this is the complete story—it doesn't explain why evolution didn't adjust the human stockpile or rate of egg depletion accordingly—it is clear that we enjoy a longer life span than our ape relatives. Why?

Some have argued that our longer lives are a very recent phenomenon, an artifact of modern medicine and more sanitary living conditions. This is misleading. It is true that life expectancy at birth has doubled in the last two hundred years. In the early 1800s, it was under forty years old, and now it is close to eighty in many places in the world. But this is primarily due to an increase in *childhood* survival. Without sanitation, basic medicine, and vaccines, the mortality rate for children used to be very high—many babies died before they were one. When you include many short lives into the life expectancy average, they drive it way down. However, in all human populations, even those with high mortality rates and those without access to medicine or sanitation, individuals who reach adulthood have a good chance of living into their sixties or beyond. Studies of hunter-gatherer populations around the globe show that about 60 percent of newborns survive to age fifteen, about 60 percent of those who reach fifteen survive until forty-five, and those who reach forty-five on average live twenty more years. In other words, it's not just a few women who reach menopause—a significant proportion of adult females in these populations are postmenopausal. This indicates that a longer life span, in particular a prolonged life after menopause, is not just a postindustrialization phenomenon (although modern environments certainly have extended our lives even longer than before, and more of us are reaching these older ages).

There is also evidence that humans have been living longer lives for thousands, hundreds of thousands, and perhaps even millions of years. Ancient Greek and Roman physicians accurately documented the age of menopause at fifty, suggesting that many women at the time were living well beyond this age. Some have looked at the fossil record to understand when human longevity evolved. Statistical models indicate that *Homo erectus* had a life span of over sixty years. An analysis of human teeth from fossils spanning the last 3 million years suggests that longevity increased throughout the course of human evolution, with the greatest jump occurring around 30,000 years ago. This finding has been questioned because these kinds of inferences from fossils are difficult to make— it is challenging to accurately assign ages to older individuals, and there are inherent biases in the fossil record itself. Nevertheless, this work collectively suggests that longevity is not just a phenomenon of the last few hundred years (although it remains to be determined just when it evolved). But this all begs the question: How does a longer life span evolve in the first place?

It goes without saying that some organisms have a very long life and some a very short one. Humans live on the order of a hundred years, whereas mice live on the order of one. How do these differences evolve? The total life span of an organism is influenced by both internal and external factors. The external factors are obvious: Your life can be cut short if you get in a car accident, if you are infected with a lethal virus, or if you are eaten by a predator. Internally, we have a clock of sorts that influences how slowly or quickly we age. This internal aging clock is set by our genes, and it can evolve just like other traits referenced in this book. For example, organisms have many ways to repair DNA and cellular damage, and a long-lived species has better and longer-functioning repair systems than a short-lived one.

There is an important link between the internal and

external factors influencing life span. In species in which there is a high chance of dying from external factors, such as mice, which are a favorite food of many carnivorous mammals and birds of prey, the internal clock tells the body to develop and age quickly. If your chance of dying young is high, the body evolves to mature quickly in order for reproduction to take place before death. A lab mouse is protected from many of the external causes of death in the wild—it obviously won't starve or get eaten by a hungry falcon in its cushy cage—but it still has a very short life, dying of "old age" between one to two years because of the aging clock it inherited from its wild mice ancestors. On the other hand, if your chance of dying from external factors is low, as in humans, your genes set a slower pace of aging. We take our time maturing, growing larger, and aging, which translates into higher reproductive output once we are ready to reproduce. As species evolve a pace of internal aging that reflects how quickly they typically die from external factors, ancestral humans must have been dying less from external factors, and their internal aging clock was adjusted accordingly.

Even for long-lived species, though, internal aging of the body cannot be avoided. If you are lucky enough to avoid external causes of death, you still have a maximum life span of about 120 years. For those who daydream about immortality, evolutionary theory clearly explains why organisms age and die, something we covered in the last chapter. In most species, the decline in performance with age affects all bodily systems at the same rate. Although a geriatric chimp in her forties still has menstrual cycles, she is not as fertile, strong, or sprightly as she was in her teens. Her internal aging clock affects her reproductive system and her other systems at about the same rate. But in species that experience menopause in midlife, like ours, the reproductive system has a faster clock than the rest of the body—the reproductive system ages more quickly than the heart, brain, or skin. Importantly, though, these systems are interconnected. Once the ovaries stop making sex hormones, all the organs with which the ovaries communicate via these hormones are affected, as I'll explain in the next section.

The menopause transition across time and space

I'm not there yet, but I'll never forget when my mother was in her late forties, a time when the freezer door was often open and there was a hand fan in every room in the house. Any woman who has been through menopause understands that it is not just about the end of reproduction. For many, the menopause transition is accompanied by a host of stressful and bothersome symptoms, including hot flashes, sweating, insomnia, vaginal dryness, fatigue, headaches, moodiness, depression, and weight gain. More seriously, hormonal changes at menopause increase a woman's risk of osteoporosis, cardiovascular disease, and dementia. But like other traits, such as menstruation, the menopause experience of women today is likely different from that of women in our evolutionary past. Let's look at the biology of the transition and then how women across the globe experience menopause and how it might have been experienced by our human ancestors.

Our ovaries don't just quit one day. In fact, they never completely quit; they do continue making tiny amounts of hormones after menstrual cycles cease. The transition to menopause is a gradual process that lasts years. To understand the transition, let's first remind ourselves of what a "normal" twenty-eight-day menstrual cycle looks like.[1] In the first two weeks of the cycle, hormones from the brain—follicle-stimulating hormone (FSH) and luteinizing hormone (LH)—stimulate the ovaries to develop dozens of follicles. One becomes dominant, pumping out huge amounts of estrogen. The estrogen signals back to the brain, inhibiting the release of FSH and inducing a surge of LH, which signals to the dominant follicle to complete development and to release its egg from the ovary around day fourteen. After ovulation, the ruptured ovarian follicle makes progesterone, preparing the uterus for a potential pregnancy.

As we enter our late thirties and forties, the number of primordial follicles left in the ovaries and the quality of the remaining follicles are vastly reduced compared to the peak.

---

1   Remember from chapter 3 that cycles are quite variable in women, including their total length and when ovulation occurs within the cycle.

There are fewer primordial follicles being woken up and fewer awakened follicles being recruited to grow in response to FSH. This influences the amounts of reproductive hormones that are produced and circulating through the body. Fewer stimulated follicles in the ovary translate to lower levels of estrogen, which in turn can lead to the release of *more* FSH, because the brain isn't getting the signal to stop releasing it. In essence, the brain starts yelling at the ovaries to develop more follicles. In some cycles, the ovaries might listen; in others, they don't. So the early stage of the menopause transition is characterized by erratic hormone levels, varying cycle lengths, and changes in the length and flow of your periods. Some women also start experiencing symptoms such as hot flashes, night sweats, insomnia, and headaches. Later during the transition, there are even fewer follicles left. Cycle lengths get very long because ovulation happens less frequently. If the ovary isn't appropriately stimulated by FSH and LH, ovulation doesn't occur, progesterone isn't produced by the ovary, the uterine endometrium doesn't transform, and so there is no menstrual period. Symptoms may become more severe.

Medically speaking, menopause has occurred after twelve months without a period. At this point, a woman's estrogen levels are low and her FSH and LH levels are high. Because these reproductive hormones also signal to other parts of the body, including the brain, bones, and fat tissue, altered levels have been linked to neurological, skeletal, and cardiovascular changes after menopause. As we are focused here on the evolutionary significance of menopause, I refer interested readers to any of several excellent books on the effects of menopause on brain and body health, including *The XX Brain* by Lisa Mosconi and *The Menopause Manifesto* by Jen Gunter.

The timing of menopause and the specific experiences of women during the menopause transition vary. Age at natural menopause varies from about forty to sixty across the world, with an average of fifty-one. As with the age at puberty, many factors influence the age at menopause, including genetics, socioeconomics, environment, and lifestyle. Family studies tell us that our genes play the largest role in menopause timing; the age a mother experiences menopause is a pretty good predictor of when her daughter will. One set of genes that influence menopause timing are those involved in repairing DNA damage. The DNA of our long-lived eggs becomes damaged for many reasons throughout our lives. Research suggests that women who repair DNA more effectively lose fewer eggs to atresia (since repaired eggs will pass the quality control filter), reaching menopause at a later age.

Nongenetic factors such as socioeconomics and lifestyle are also important, although the associations are less clear. The age an older sister experiences menopause is an even better predictor of your own age at menopause than your mother's, because sisters share genes *and* more of the nongenetic factors that influence menopause timing. Smoking has the most consistent association—menopause occurs earlier in smokers and in women whose mothers smoked during pregnancy, likely because tobacco toxins kill eggs. Other factors that appear to be associated with an earlier age of menopause include childhood undernutrition, poor health, low education levels, and low socioeconomic status, although many of these factors are related so it is difficult to tease apart which have stronger and/or independent effects. On average, women from Africa, the Middle East, and some Latin American and Asian countries experience menopause earlier than those from Europe, Australia, Canada, and the United States, but it isn't known if these differences are caused by genetic or nongenetic factors or both. Women who have delivered more children and who have their last child at an older age experience menopause later than those with no children, but it isn't clear if the process of pregnancy, birth, and lactation directly affects menopause timing or simply that women with more children have a healthier follicle supply to begin with. Some studies have shown that taking oral contraceptives is associated with a later age at menopause, but others have found no such correlation. Although the pill and pregnancy/lactation stop your ovaries from releasing an egg each cycle, they don't stop the primordial follicles that are continuously being awakened from developing and then dying of atresia.

Some of the same factors that influence the timing of menopause also influence the symptoms experienced during the menopause transition. Cross-cultural studies have shown that the symptoms of menopause, like hot flashes, are not universal. Women from Western societies report more stress and more symptoms around the menopause transition than those from other populations. These differences may be the result of differences in our underlying genetics, in environments/lifestyles, and/or in cultural attitudes about menopause. A gene variant on chromosome 4 is associated with a higher risk of hot flashes. Environment and lifestyle factors also play a role; smokers and women with higher BMI are likely to experience more, and more severe, hot flashes.

In thinking about when and how our ancestors experienced menopause, it is useful to remember what their lives were like. Their diets and levels of physical activity were drastically different from those of many women today, and ancestral women spent much of their adult life pregnant or nursing. Their accumulated exposure to ovarian hormones was lower than it is today, since women who are pregnant and nursing are not cycling. In addition, high-calorie, high-fat diets in some populations today, combined with low levels of physical activity, are associated with higher estrogen and progesterone levels during each menstrual cycle. As anthropologist Wenda Trevathan has argued, higher estrogen and progesterone levels are associated with more severe PMS symptoms, possibly because of a steeper drop of these ovarian hormones at the end of each menstrual cycle. By similar logic and also proposed by Trevathan (and others), the ultimate withdrawal of these hormones at menopause after decades of high levels might lead to more severe menopause symptoms in populations with diets and lifestyles that are so vastly different from those of our ancestors.

Lifestyle factors during the transition itself likely also influenced the way our ancestors experienced menopause. A woman in the past probably nursed her last child for a longer period of time than her other children, perhaps into her midforties, and then never resumed ovulation, as has been observed in some hunter-gatherer populations. High levels of prolactin and oxytocin, those feel-good mothering hormones, potentially alleviated the negative effects of declining estrogen and progesterone. Lifestyle factors after menopause probably also influenced a woman's health and quality of life. A study on Mayan women found that although women do experience a decline in estrogen and a decrease in bone mineral density after menopause, these women do not experience more fractures, perhaps because of higher levels of physical activity and healthier diets.

———

Both the age at menopause and the symptoms experienced at menopause vary across women and are influenced by a complex interplay of genetics, environment, lifestyle, and culture. Because life was drastically different for women in our evolutionary past, in particular their diets and levels of physical activity, they likely did not experience menopause the way that we do today.

Why did menopause and longevity evolve?

Some women reach menopause at forty, some at fifty, and some at fifty-five or even sixty. This variation is largely determined by genetics, and genetic variation is one of the requirements for evolutionary change. Assuming that the age at menopause also varied in our human ancestors, the pressing question is why our reproductive spans didn't lengthen over time in parallel with our life spans. Why hasn't menopause shifted into the late fifties, sixties, or seventies? In theory, the evolutionary solution is simple: Make more eggs up front, or slow down the rate of follicle depletion. In practice, some long-lived species appear to have done this, including elephants, baleen whales, and naked mole rats, as mentioned above. Why not humans, killer whales, narwhals, beluga whales, or short-finned pilot whales?

We don't have an answer to this question yet, and the answer might not be the same for each of these species.[2] But the fact that a long life after menopause is observed in some whales indicates that midlife menopause can evolve under natural living conditions. Obviously killer whales aren't getting a yearly flu shot or taking meds for high cholesterol, and human longevity existed long before twentieth-century improvements in health care and sanitation.

---

2   Beluga whales and narwhals are closely related species, so a long life after menopause probably evolved once in their ancestor. The other species are more distantly related, and menopause/longevity likely evolved independently in these species, potentially for different reasons.

The best-known evolutionary explanation for human menopause—the grandmother hypothesis—doesn't actually explain why we experience menopause but rather why we live longer. According to the hypothesis, we inherited menopause at fifty from our ape ancestors, and our longevity is the trait that needs explaining. The earliest mention of the grandmother hypothesis was in the 1950s, but anthropologist Kristen Hawkes further developed the idea in the 1990s after her work with a tribe of hunter-gatherers from East Africa called the Hadza. The Hadza live in multigenerational family groups, and Hawkes noticed that postmenopausal Hadza women play an important role in their families. They dig and process tubers, which compose a large part of the Hadza diet and are foods that young children can't manage on their own. Since most young Hadza couples live with the woman's family, grandmothers are often provisioning their daughters' children. The help a grandmother provides boosts her reproductive success by boosting that of her daughters. A helpful Hadza grandmother has more grandchildren.

Based on these observations, Hawkes put forth a hypothesis about the evolution of human longevity that centered on maternal grandmothers. The hypothesis predicts ecological pressures at some point in the past (as long as 1.8 million years to as recently as 50,000 years ago) that restricted food availability for young children in particular. Longer-lived females who helped feed their weaning grandchildren left more descendants than those who didn't. The genes of these robust and cooperative grandmothers—in particular, the genes involved in their longevity and generous behavior—were propagated in future generations because more grandchildren were surviving and passing on these same genes. Over time, this led to the evolution of even longer lives, shorter intervals between births, and longer childhoods, unique features of human life history. The grandmother hypothesis is an evolutionary explanation for human longevity based on the concept of kin selection. Even though postmenopausal women in the past weren't reproducing themselves, they boosted their reproductive success by improving the survival of their grandchildren.

The grandmother hypothesis turned some standard assumptions in the field of human evolution on their head. It has often been argued that longer childhoods evolved in humans to allow for the development of a larger brain and all the learning required to be a human adult. Moreover, according to old assumptions, an increase in meat consumption when men started hunting large game facilitated the evolution of these longer childhoods and bigger brains. Instead, Hawkes argued that longer childhoods evolved because longer lives evolved—in the animal world, a long life usually predicts a long childhood—and that longer lives evolved because of helpful grandmothers. In Hawkes's vision of human evolution, helpful grandmothers came first.

Another standard assumption in human evolutionary studies is that our female ancestors left their family groups when they hit puberty to join new ones. In most social mammals and primates, males leave the group, probably to avoid inbreeding, but in apes, it is often the males that stay and the females that leave. Many models of human evolution are built with female dispersal from the group as a given. But the grandmother hypothesis works best—both in theory and in studies of real populations—when mothers and daughters stay together and mothers provide for their daughters' children. Female dispersal is not necessarily a deal breaker for the hypothesis. If daughters leave and mothers provide for their sons' children instead, they can certainly boost their reproductive success, but the uncertainty of paternity will dilute the effects of their generous behavior. If a woman stayed together with one man but had children with many, it is less likely that helpful behavior on the part of a grandmother would have translated into more grandchildren because she might have been helping someone else's.

One big unresolved problem in the field is who left and who stayed in ancestral human groups. Because we often have to guess about past behaviors based on what we observe in the present and preindustrial human and ape societies today are characterized by both female and male dispersal at maturity, this issue may never be resolved.

Since that work on the Hadza, other hunter-gatherer and preindustrial data sets from places like Costa Rica, Finland, and Gambia have been examined, with mixed support for the hypothesis. Some of the inconsistencies likely stem from different patterns of dispersal in these human groups. Much work has also been done on killer whales, which are social, group together in extended family networks, and have a long postmenopausal life stage like humans. As the work on the Hadza showed, a grandmother's presence in the group increases survival of her grandchildren. Apparently, grandma killer whales in pods that feast on salmon are the experts at finding salmon, which is particularly useful in years of low salmon abundance.

So there is a good deal of evidence that in certain contexts, grandmothers improve the survival of their grandchildren. But does that provide a sufficient explanation for the evolution of menopause and longevity? There are a couple of issues with the grandmother hypothesis that we need to address. The first concerns the order of events, and the second concerns a key omission in the hypothesis—why women didn't evolve a later menopausal age as life span lengthened.

Anthropologists Peter Ellison and Mary Ann Ottinger object to the order of trait evolution under the grandmother hypothesis. The crux of their criticism is that helpful behavior by postmenopausal grandmothers can't explain the existence of postmenopausal grandmothers capable of doing the helping in the first place. Natural selection works better with bigger numbers; evolution needs many, many opportunities to make an adaptation precise. Ellison and Ottinger argue that humans must have started living longer for some other reason, and only then were there enough postmenopausal women around for grandmotherly behavior to evolve. It's like the classic riddle of the chicken and the egg. Which came first: longer lives or indulgent grandmothers?

Ellison and Ottinger make an interesting connection between humans and captive or domestic animals, such as lab mice, monkeys at a primate facility, chickens at a free-range poultry farm, or breeding minks in a *Golden Girls* episode. These females often live beyond their reproductive span because of decreased mortality rates in captivity—they aren't dying of starvation or predation as their wild counterparts do.[3] In Ellison and Ottinger's view, adult mortality rates went down in ancient human groups with the emergence of cooperation, food sharing, and a division of labor, a process described as self-domestication. As a result, many human adults started living beyond the age of menopause, like captive minks. Unlike captive minks, however, which continue to be fed and cared for by their human sponsors until death, postmenopausal women in the past were potentially diverting resources from younger members of the group who were still reproducing.

Ellison and Ottinger envision two possible evolutionary solutions to this problem. First, women could realign their reproductive span to their now longer life span by making more eggs and/or slowing down their rate of egg depletion. Second, they could alter their behavior to benefit younger members of the

---

3   The fact that some captive animals do live beyond the age when they run out of eggs has led to some confusion about how widespread menopause is in the animal world. However, as we discussed earlier in the chapter, life span and egg stockpile sizes evolve in response to typical mortality rates in a species. If you remove those causes of external mortality, an animal will live longer than its wild counterpart, although it usually doesn't represent a significant proportion of life span.

group, which would indirectly benefit themselves if the recipients were close relatives. They argue that option two—cooperative behavior on the part of postmenopausal grandmothers—was taken because it was the easiest and quickest solution. Cooperative behavior was already part of the human behavioral repertoire, and it would have quickly resulted in evolutionary benefits to postmenopausal grandmothers. Although option one would have yielded higher reproductive success for aging females—you pass on twice as many genes through your children as through your grandchildren—the path to a longer reproductive life span was harder to take.

Other biologists have also argued that a later age at menopause didn't evolve because it was too difficult or risky. Perhaps increasing the size of the egg stockpile would have been too onerous and slow—it has been estimated that a woman would need to make millions more follicles to delay menopause by just three or four years. Perhaps slowing the rate of follicle depletion would have been too big a risk for our ancestors—waking up fewer follicles each month might reduce fertility early in life, and relaxing quality control would increase the risk of genetic abnormalities in children. Some have suggested that eggs can't last longer than forty-five to fifty years, that they have a "shelf life." Finally, even if ancestral women had evolved a larger/longer-lasting egg supply, perhaps they wouldn't have been able to support a pregnancy or manage childbirth and childcare in their late forties or fifties. Childbirth is especially hard on human women who give birth to such large-brained babies, and the risk of mother (and baby) dying during childbirth goes up with increasing maternal age.

All that said, some species have managed to solve the problem of lengthening reproductive span without being held back by these issues, so I'm not yet convinced that it wouldn't have been possible for our species. Naked mole rats have a huge egg supply, and the eggs of baleen whales have a centennial shelf life. Perhaps relaxing quality control is not a feasible evolutionary option, but what about improving DNA repair to keep the egg supply healthier for longer? (Women with later menopause appear to have enhanced DNA repair, as mentioned earlier.) Even the risk of death during childbirth can't fully explain why menopause must happen so early in humans. Childbirth is a unique challenge in our species, but in models and analyses of human data, it appears the risk of maternal death isn't high enough, nor are the costs—in the historical populations studied, babies whose mothers died in childbirth survived with the help of other family members.

Yet another take on why reproductive span was kept short in our female ancestors is based on the idea of genetic conflict. We've covered many conflicts in this book—between a mother and fetus over resources, between female and male over reproduction, between female and male over how much to invest in children. Let's add another duo to the list of warring parties— between mother-in-law and daughter-in-law. A discussion of conflict wouldn't be complete without including the in-laws!

The central premise of the "reproductive conflict" hypothesis is that conflict over scarce resources was resolved when older women stopped reproducing around the time younger women started. A pregnant and nursing woman needs hundreds of extra calories per day; in ancestral human groups that pooled and shared resources, there were only so many calories to go around. How hard a mother was willing to fight for those resources depended in part on who she was taking them away from. In contrast to the grandmother hypothesis, the reproductive conflict hypothesis works best when human females leave the group at reproductive maturity. When a female joins a new group, she isn't related to anyone until she starts having children, so she's diverting resources for her children away from people she isn't related to and is willing to fight pretty hard for those calories. In contrast, the "mother-in-law"[4] is related to her own children *and* the daughter-in-law's children—they're her grandchildren—so if the mother-in-law continues having children, she is diverting resources from her own relatives. Under this scenario, mothers-in-law face disproportionate costs when fighting for those resources. One solution is for them to surrender and stop reproducing. Menopause resolves the conflict.

The reproductive conflict hypothesis is supported by the observation that in species with midlife menopause, there is little overlap of reproduction between generations. When a

---

4   I use the term *in-law* here because it is familiar, but laws and marriage did not exist in our hunter-gatherer ancestors.

woman starts having children, her mother has finished. When a female killer whale starts having children, her mother has finished. Although rare in killer whales, in cases where older and younger females breed at the same time, the calves of older females die at a higher rate than those of younger ones, providing support for the hypothesis.[5] On the other hand, studies of preindustrial human groups provide mixed support for the hypothesis, likely in part because of different dispersal patterns in these groups.

Remarkably, in some mammalian species, in particular those that communally raise offspring, like mongooses, hyenas, and naked mole rats, *younger* females of the group are those whose reproduction is suppressed. Reproductive suppression in young females is usually reversible (unlike menopause in older females), but it is nevertheless an evolved trait that is a necessary feature of group life in these species. Young, suppressed females are helpers (as are postmenopausal Hadza grandmothers), enabling the high reproductive output of older females by feeding them during pregnancy and lactation and feeding/ caring for their children. Reproductive suppression in young or old females is not a widespread phenomenon in vertebrates, but where it exists, an important variable influencing who forgoes

---

5   Killer whales have different dispersal patterns than other mammals. Both males and females mate with nongroup members but then come back to live in their group. So mother killer whales are in conflict with their own daughters, not the mates of their sons, as predicted in ancestral humans.

reproduction is the level of relatedness of group members of different ages to other members of the group, which in turn is influenced by dispersal patterns. If, indeed, ancestral human females left the group at maturity and there was conflict between these young females and the mothers of their mates over resources for reproduction, the evolutionary strategy that worked best for aging females may have been to stop reproduction at menopause and invest in grandchildren.

While it has long been assumed that we inherited menopause at fifty from our ape ancestors, a recent comparison of follicle depletion rates in humans and chimps suggests that we actually deplete our egg supply much faster than chimps. By comparing the number of primordial follicles present in the ovaries of humans and chimps at different ages, a team led by Kristen Hawkes (of the grandmother hypothesis) found that the rate of follicle depletion in humans and chimps is similar up to age thirty-five but that humans lose follicles much faster than chimps after that age. If human egg depletion continued at the original, slower rate, we'd have enough eggs to last until we're over seventy. Clearly, more research is needed here, starting with a larger sample size, in particular including more chimps over thirty-five (which aren't easy to come by). Studying the underlying cause of the difference would be another fruitful avenue of research—the authors of the study suggest it lies in the brain, in particular the regions that produce and release the hormones that communicate with the ovaries. If the observation of a faster follicle depletion rate in women over thirty-five held up, it would suggest that reproductive conflict between generations (or another, as yet unknown force) didn't just hold back a longer reproductive span in women but also shortened it. Don't underestimate the power of your daughter-in-law!

───

After this foray into the evolutionary theories on menopause, what can we conclude about the purpose of midlife menopause in our species?

I discussed the fact that all mammals make a finite supply of eggs during fetal development, which means that a female mammal will run out of eggs if she lives long enough. But in most species, the supply has been adjusted by natural selection to last until the end of natural life. Therefore, early menopause is rare in the animal world, observed in humans and a few species of toothed whales. The reasons are not yet clear. We know that humans and those toothed whale species have longer natural life spans than closely related species and that humans and killer whales live in cooperative family groups. We also know that grandmothers can increase the survival of their grandchildren in certain contexts and that there is potential conflict between older and younger females over reproduction. None of this is unique to species with menopause. In the social and long-lived elephants, for example, grandmothers are known to boost the survival of their grandchildren, but they don't experience menopause. Why not?

When we're making personal decisions in life, such as which job to take or whether to have a child, we often compare the costs to the benefits. The cost-benefit ratio depends on the context; for example, it might be much higher for a tenth child than for a first. When you have that first child, you might really appreciate having your mother move in to take care of your newborn while you go to work, but when the child is old enough to go to school, the cost-benefit ratio may have changed. Similarly, the evolutionary "decision" of midlife menopause in our species was likely influenced by the specific social, ecological, and demographic context in which our ancestors lived. As life span lengthened in our species, the benefits of postmenopausal helping, together with the costs of extending reproduction, likely explain why midlife menopause exists in our species but not in most.

One reason we don't have a solid evolutionary explanation for menopause is that we have such a poor understanding of its mechanistic underpinnings. Half of the human population experiences menopause, but the data and research spending on women's health and aging have been woefully lacking. We don't understand the role of the brain versus the ovaries in the transition to menopause. We don't understand why we start losing our follicles so quickly in our thirties and forties. We don't have a good diagnostic tool that can predict a woman's fertility status. We aren't close to developing therapeutics that might delay the onset of menopause and associated symptoms (while menopause likely evolved for good reasons, the benefits don't necessarily carry over to the modern world, leaving us with another mismatch between our evolved biology and modern lives). Answering these kinds of questions is critically important for improving women's reproductive and overall health. And as ovarian aging also affects our family planning and career choices, it is a women's equality issue as well.

Fortunately, the research tides are slowly starting to change. Places like the Buck Institute for Research on Aging—my current professional home—are starting ambitious research centers and programs focused on understanding how the female reproductive system ages and how ovarian aging affects the rest of the body. Investors have also realized that menopause treatment is a huge untapped market. Telemedicine companies like Evernow are conducting large-scale research studies on menopause symptoms, which will help providers create more personalized treatment options for women. It looks like grandmothers (and their daughters and granddaughters) are finally starting to get the attention they deserve.

In the meantime, with the help of their two adoring grandmothers, I've been able to complete this book while chasing after my four children, crystallizing for me the importance of grandmothers in my lifetime and in past lifetimes as well. Although my own grandmother, who's in her nineties, isn't able to help me with my kids, I remember the days when she helped my parents with me. During regular sleepovers at her house, she'd make my brother and me curly pasta with sauce

CONCLUSION

Lessons of the Past

Modern female biology is an amazing collection of the old and the new. Most of our traits are ancient, revealing the deep connections we have with other life on the planet. We share the basics of our reproductive cycle with all vertebrates, mammary glands and lactation with all mammals, placentas and pregnancy with all placental mammals, and menstruation with all monkeys and apes. Each of these traits has its own remarkable story of how and why it came to be. While they aren't ours alone, these traits (and many others) are the foundation of our biology, setting the stage for its unique evolution in humans. From my perspective—I am an evolutionary biologist after all—these ancient parts of us are fascinating and beautiful in their own way.

Other traits are younger, emerging after humans and chimpanzees split from a common ancestor over 6 million years ago. Many of these were inherited from our common ancestor but were modified in humans. Humans evolved an efficient, exclusive mode of upright locomotion. Our brains and bodies became larger, while the size difference between males and females became smaller. We have advanced intelligence, social cognition, language, and culture. We have more kids (at least in communities that don't use birth control). We wean our children early, and our childhoods are especially long. We live in cooperative groups with a division of labor. We live longer, and we experience menopause in midlife.

Working out human evolutionary history after our split from chimpanzees reminds me of solving those problems in fourth grade math on the order of operations. You're given an equation, for example, $7-1+55 \div 5=x$, and you need to figure out the order to perform the arithmetic to solve it correctly. For those who know the rules (please excuse my dear aunt Sally!), the problem is straightforward, but for those who are just learning, it's a challenge. In the case of human evolution, we already know the answer on the right side of the equation—human biology today—and we know many of the events on the left side of the equation that must have occurred since we split from a common ancestor with chimpanzees. What we don't know is the precise order of events that lead us to the answer, and there aren't any convenient rules or mnemonics to help us out. Clearly, many human traits are connected and depend on the presence of the other ones, leaving us wondering, like in the riddle of the chicken and the egg, which came first.

We don't know the precise order in which these human traits evolved, but we can be sure that most did not evolve sequentially, one after the other, but rather together, interacting with each other as they changed. Maximum life span didn't jump from 60 to 120 all at once. Our brains didn't go from small to large overnight. Even a trait as seemingly straightforward as upright locomotion did not evolve all at once but more likely in stages. We need to introduce a new operation into the human evolution equation, one that generates feedback and reinforcement between traits. A slightly longer life reinforces a slightly longer childhood reinforces a slightly larger brain reinforces more social cooperation reinforces a slightly longer life, and so on. Human evolution is complicated, with both positive and negative feedback loops connecting and reinforcing many of the traits that define our species.

Menopause, longevity, long childhoods, large brains, early weaning, and social cooperation are part of a suite of connected traits that contributed to the success of our ancestors and ultimately the success of our species. While our cultural sense of worth and success today is not measured only by how many children or grandchildren we have, the evolutionary success of our ancestors was. The traits and behaviors that allowed some of our ancestors to leave more descendants than their peers are those that persisted and spread. The human brain evolved a level of complexity and sophistication not observed in other animals, to the point that we can now decide how to measure our own success, whether it's raising a family, creating a new technology, leading a company, finding serenity, or writing a bestselling book.

Some of the traits we've discussed in this book are best

explained as adaptations to the *physical* world. The transition to upright locomotion occurred in stages over a long period of major climate change during which the environment shifted from rain forest to grassland. Those who could stand, walk, and run efficiently on two feet were better foragers and hunters in this new physical world. The evolution of pregnancy and live birth, which occurred multiple times in vertebrates, is likely an adaptation that better protects young offspring from ecological challenges such as predators or the cold.

Many of the traits we've discussed, however, emerged as adaptations in an ever-changing *social* world. Although pregnancy may have first evolved in mammals as a solution to harsh temperatures or predation, it continued to evolve in the ongoing genetic negotiations between mother and fetus over resources. These negotiations are constrained by but operate somewhat independently of the physical environment. Likewise, the mammary gland may have evolved to reliably feed young offspring in an unpredictable physical world, but further changes in humans—the permanent, fatty breast— likely took place as a result of ongoing social exchanges with potential mates. While the origins of human menopause, longevity, and grandmothering will continue to be debated for years to come, cooperation and conflict between the genes of family members likely contributed to the evolution of these traits. In a social species like ours, both ecological and social factors have shaped our biology, including our outward appearance, physiology, behavior, and life history.

As I navigate my own relationships in a modern social and physical world, such as with family, my larger community, food, and exercise, I try to think about the relationships of our ancestors that molded our biology to what it is today. When I'm exasperated with my kids for incessant bickering, annoyed with my husband for not helping around the house, disappointed in myself for bingeing on potato chips, or frustrated with my body for causing me so much pain during a period, I pause and remind myself of the long evolutionary history of these traits. I won't pretend that these reminders have drastically changed my life. While I do try to eat balanced meals, stay active, and avoid smoking and heavy drinking, I am not a dieter, a marathon runner, or the most patient mother or partner. But understanding why my kids bicker and why those potato chips taste so good allows me to be a little more generous with myself and with others.

Studying our evolutionary history also reminds me of the flexibility that is built into our systems. Our genes do not define us or decide our fates; their actions are also influenced by factors that are under our control, such as what we eat, with whom we interact, and how much we move our bodies. The complex human brain is especially flexible. Yes, some of our behaviors have deep roots, but even these ancient behaviors are flexible, such as in the context of mothering. Our flexible and creative brains have allowed us to colonize almost every habitat on the planet and to develop cultures that produce extraordinary literature, art, medicine, architecture, and technology. Human culture allows us to deal with and even rise above some of our undesirable biological legacies. We can take Advil for our menstrual cramps or avoid periods altogether; we can educate ourselves and others about the benefits of healthy diets and exercise; we can cure some cancers and take vaccines to protect against infections. I hope we can embrace this flexibility of mind and body in the future to forge healthier relationships with ourselves, each other, and our planet.

Acknowledgments

My book journey began many years ago when I was a graduate student at Yale, where I started work on the female reproductive system with Günter Wagner. I still feel fortunate to have been mentored by Günter, an exceptional biologist who approaches his work with an inspiring combination of curiosity, creativity, and diligence. Many other scientists have supported and inspired me along the way, including Susan Antón, Todd Disotell, Stephen Stearns, Mihaela Pavlicev, Vivian Irish, Jeff Townsend, Hugh Taylor, Paul Turner, Vincent Lynch, Rebecca Young, James Noonan, and Craig Miller. At the Buck Institute, I'd like to acknowledge Jennifer Garrison, Eric Verdin, Chris Benz, and Malene Hansen for their support. This book would not exist without Jennifer, a gifted community builder and visionary in the field of female reproductive aging. Thank you for your encouragement, generosity, and friendship, Jennifer.

As I moved to turn my idea into an actual book, it started with Gina Pell, one of the coolest and smartest women I know.

In the time I've known Gina, she and the wonderful Amy Parker started a network for women, The What Alliance, where I've met exceptional and supportive women both online and in person. Through Gina I connected with Jackie Ashton, my talented and savvy literary agent at Lucinda Literary. Jackie helped me find the perfect home for my book, Sourcebooks, and a passionate editor, Erin McClary. Erin had the vision to see what my book could become and the skill to move each draft closer to that vision. Many thanks to everyone at Sourcebooks involved in editing, producing, and selling my book, including Findlay McCarthy, Sabrina Baskey, Emily Proano, Brittney Mmutle, and Madeleine Brown. I am also grateful to colleagues who commented on chapter drafts: Stephen Stearns, Günter Wagner, Mihaela Pavlicev, Becca Young, Mike Ryan (and his lab members), and Shana Caro. I am especially thankful to Steve for providing valuable feedback on multiple drafts.

Catherine Cox, Aparna Kollipara, Nuria Kruger, Haleh Partovi, Brenda Scheumann, and Shanthi Sekeran have been dear friends and cheerleaders. Thanks to Zinaida Zybach for thoughtful conversations about my book and help with my website. I am eternally grateful to my family for sustaining me with love and support. I appreciate my mother, Randa, for her encouragement and incisive way of looking at the world, and my father, Tarek, whose belief in me will stay with me forever. Thank you to my in-laws, Afaf and Mahmoud Meleis, for always lifting me up with positivity and help on the ground. I am most grateful to my husband, Sherief Meleis, whose support behind the scenes was crucial as I wrote this book. We have a hectic schedule as parents of four athletic and social children, but any time I come to him with an idea, whether it's something small like needing to take a dinner break with friends or something huge like writing a book, he is invariably positive, encouraging, and generous. When I first floated the idea of writing this book, he said, "Great idea— you should do it!" Thanks to you, Sherief, here it is.

Bibliography

Abbot, Patrick, and John A Capra. "What Is a Placental Mammal Anyway?" *eLife* 6 (September 2017): e30994. https://doi.org/10.7554/eLife.30994.

Adams, Daniel B. "The Cheetah: Native American." *Science* 205, no. 4411 (September 1979): 1155–58. https://doi.org/10.1126/science.205.4411.1155.

Allen, M. L., and W. B. Lemmon. "Orgasm in Female Primates." *American Journal of Primatology* 1, no. 1 (1981): 15–34. https://doi.org/10.1002/ajp.1350010104.

Allison, A. C. "Protection Afforded by Sickle-Cell Trait against Subtertian Malarial Infection." *British Medical Journal* 1, no. 4857 (February 1954): 290–94. https://doi.org/10.1136/bmj.1.4857.290.

Alonso, Laura C., and Robert L. Rosenfield. "Oestrogens and Puberty." *Best Practice & Research Clinical Endocrinology & Metabolism* 16, no. 1 (March 2002): 13–30. https://doi.org/10.1053/beem.2002.0177.

American Academy of Pediatrics, Committee on Adolescence, American College of Obstetricians and Gynecologists, and Committee on Adolescent Health Care. "Menstruation in Girls and Adolescents: Using the Menstrual Cycle as a Vital Sign." *Pediatrics* 118, no. 5 (November 2006): 2245–50. https://doi.org/10.1542/peds.2006-2481.

American Cancer Society, "Cancer Facts & Figures 2022," cancer.org, 2022, https://www.cancer.org/research/cancer-facts-statistics/all-cancer-facts-figures/cancer-facts-figures-2022.html.

Amundsen, Darrel W., and Carol Jean Diers. "The Age of Menopause in Classical Greece and Rome." *Human Biology* 42, no. 1 (February 1970): 79–86. http://www.jstor.org/stable/41449006.

Andersson, Malte. "Female Choice Selects for Extreme Tail Length in a Widowbird." *Nature* 299, no. 5886 (1982): 818–20. https://doi.org /10.1038/299818a0.

Andreakos, Evangelos, Laurent Abel, Donald C. Vinh, Elżbieta Kaja, Beth A. Drolet, Qian Zhang, Cliona O'Farrelly, et al. "A Global Effort to Dissect the Human Genetic Basis of Resistance to SARS-CoV-2 Infection." *Nature Immunology* 23, no. 2 (February 2022): 159–64. https://doi.org/10.1038/s41590-021-01030-z.

Arboleda, Valerie A., David E. Sandberg, and Eric Vilain. "DSDs: Genetics, Underlying Pathologies and Psychosexual Differentiation." *Nature Reviews Endocrinology* 10, no. 10 (October 2014): 603–15. https:// doi.org/10.1038/nrendo.2014.130.

Bachtrog, Doris, Judith E. Mank, Catherine L. Peichel, Mark Kirkpatrick, Sarah P. Otto, Tia-Lynn Ashman, Matthew W. Hahn, et al. "Sex Determination: Why So Many Ways of Doing It?" *PLoS Biology* 12, no. 7 (July 2014): e1001899. https://doi.org/10.1371/journal.pbio.1001899.

Baker, R. Robin, and Mark A. Bellis. "Human Sperm Competition: Ejaculate Manipulation by Females and a Function for the Female Orgasm." *Animal Behaviour* 46, no. 5 (November 1993): 887–909. https://doi.org/10.1006/anbe.1993.1272.

Ballard, Olivia, and Ardythe L. Morrow. "Human Milk Composition." *Pediatric Clinics of North America* 60, no. 1 (February 2013): 49–74. https://doi.org/10.1016/j.pcl.2012.10.002.

Barber, Iain, Stephen A. Arnott, Victoria A. Braithwaite, Jennifer Andrew, and Felicity A. Huntingford. "Indirect Fitness Consequences of Mate Choice in Sticklebacks: Offspring of Brighter Males Grow Slowly but Resist Parasitic Infections." *Proceedings of the Royal Society B: Biological Sciences* 268, no. 1462 (January 2001): 71–76. https:// doi.org/10.1098/rspb.2000.1331.

Barraclough, Charles A., and Roger A. Gorski. "Evidence That the Hypothalamus Is Responsible for Androgen-Induced Sterility in the Female Rat." *Endocrinology* 68, no. 1 (January 1961): 68–79. https:// doi.org/10.1210/endo-68-1-68.

Basson, Rosemary. "The Female Sexual Response: A Different Model." *Journal of Sex & Marital Therapy* 26, no. 1 (2000): 51–65. https://doi.org/10.1080/009262300278641.

Beck, Kristen L., Darren Weber, Brett S. Phinney, Jennifer T. Smilowitz, Katie Hinde, Bo Lönnerdal, Ian Korf, and Danielle G. Lemay. "Comparative Proteomics of Human and Macaque Milk Reveals Species-Specific Nutrition during Postnatal Development." *Journal of Proteome Research* 14, no. 5 (2015): 2143–57. https://doi.org/10.1021/pr501243m.

Beery, Annaliese K., and Irving Zucker. "Sex Ratio Adjustment by Sex-Specific Maternal Cannibalism in Hamsters." *Physiology & Behavior* 107, no. 3 (October 2012): 271–76. https://doi.org/10.1016/j.physbeh.2012.09.001.

Bellis, Mary. "A Brief History of the Tampon." ThoughtCo, September 8, 2021. https://thoughtco.com/history-of-the-tampon-4018968.

Bellofiore, Nadia, Stacey J. Ellery, Jared Mamrot, David W. Walker, Peter Temple-Smith, and Hayley Dickinson. "First Evidence of a Menstruating Rodent: The Spiny Mouse (*Acomys cahirinus*)." *American Journal of Obstetrics and Gynecology* 216, no. 1 (2017): 40.e1–11. https://doi.org/10.1016/j.ajog.2016.07.041.

Bellofiore, Nadia, and Jemma Evans. "Monkeys, Mice and Menses: The Bloody Anomaly of the Spiny Mouse." *Journal of Assisted Reproduction and Genetics* 36, no. 5 (2019): 811–17. https://doi.org/10.1007/s10815-018-1390-3.

Benirschke, Kurt. Comparative Placentation (website). Updated January 19, 2012. http://placentation.ucsd.edu/index.html.

Benowitz-Fredericks, Z. M., Alexander S. Kitaysky, Jorg Welcker, and Scott A. Hatch. "Effects of Food Availability on Yolk Androgen Deposition in the Black-Legged Kittiwake (*Rissa tridactyla*), a Seabird with Facultative Brood Reduction." *PLoS ONE* 8, no. 5 (2013): e62949. https://doi.org/10.1371/journal.pone.0062949.

Benton, Michael J., Danielle Dhouailly, Baoyu Jiang, and Maria McNamara. "The Early Origin of Feathers." *Trends in Ecology & Evolution* 34, no. 9 (September 2019): 856–69. https://doi.org/10.1016/j.tree.2019.04.018.

Berridge, Kent C., and Morten L. Kringelbach. "Pleasure Systems in the Brain." *Neuron* 86, no. 3 (May 2015): 646–64. https://doi.org/10.1016/j.neuron.2015.02.018.

Beydoun, H., and A. F. Saftlas. "Association of Human Leucocyte Antigen Sharing with Recurrent Spontaneous Abortions." *Tissue Antigens* 65, no. 2 (February 2005): 123–35. https://doi.org/10.1111/j.1399-0039.2005.00367.x.

Blaicher, Wibke, Doris Gruber, Christian Bieglmayer, Alex M. Blaicher, Wolfgang Knogler, and Johannes C. Huber. "The Role of Oxytocin in Relation to Female Sexual Arousal." *Gynecologic and Obstetric Investigation* 47, no. 2 (February 1999): 125–26. https://doi.org/10.1159/000010075.

Bode, Lars. "Human Milk Oligosaccharides: Every Baby Needs a Sugar Mama." *Glycobiology* 22, no. 9 (September 2012): 1147–62. https://doi.org/10.1093/glycob/cws074.

Borgia, Gerald. "Bower Quality, Number of Decorations and Mating Success of Male Satin Bowerbirds (*Ptilonorhynchus violaceus*): An Experimental Analysis." *Animal Behaviour* 33, no. 1 (February 1985): 266–71. https://doi.org/10.1016/S0003-3472(85)80140-8.

Bourke, Andrew F. G. "Hamilton's Rule and the Causes of Social Evolution." *Philosophical Transactions of the Royal Society B: Biological Sciences* 369, no. 1642 (2014): 20130362. https://doi.org/10.1098/rstb.2013.0362.

Brawand, David, Walter Wahli, and Henrik Kaessmann. "Loss of Egg Yolk Genes in Mammals and the Origin of Lactation and Placentation." *PLoS Biology* 6, no. 3 (2008): e63. https://doi.org/10.1371/journal.pbio.0060063.

Briskie, James V., Christopher T. Naugler, and Susan M. Leech. "Begging Intensity of Nestling Birds Varies with Sibling Relatedness." *Proceedings of the Royal Society B: Biological Sciences* 258, no. 1351 (1994): 73–78. https://doi.org/10.1098/rspb.1994.0144.

Brown, Jerram L., and Amy Eklund. "Kin Recognition and the Major Histocompatibility Complex: An Integrative Review." *American Naturalist* 143, no. 3 (March 1994): 435–61. https://doi.org/10.1086/285612.

Brown, Richard E., Bruce Roser, and Prim B. Singh. "Class I and Class II Regions of the Major Histocompatibility Complex Both Contribute to Individual Odors in Congenic Inbred Strains of Rats." *Behavior Genetics* 19, no. 5 (September 1989): 659–74. https://doi.org/10.1007/BF01066029.

Bryant, Gregory A., and Martie G. Haselton. "Vocal Cues of Ovulation in Human Females." *Biology Letters* 5, no. 1 (2009): 12–15. https://doi.org/10.1098/rsbl.2008.0507.

Buisson, Odile, Pierre Foldes, and Bernard-Jean Paniel. "Sonography of the Clitoris." *Journal of Sexual Medicine* 5, no. 2 (February 2008): 413–17. https://doi.org/10.1111/j.1743-6109.2007.00699.x.

Burgoyne, Paul S., and Arthur P. Arnold. "A Primer on the Use of Mouse Models for Identifying Direct Sex Chromosome Effects That Cause Sex Differences in Non-Gonadal Tissues." *Biology of Sex Differences* 7, no. 68 (December 2016): 1–21. https://doi.org/10.1186/s13293-016-0115-5.

Buss, David M. "Sex Differences in Human Mate Preferences: Evolutionary Hypotheses Tested in 37 Cultures." *Behavioral and Brain Sciences* 12, no. 1 (1989): 1–14. https://doi.org/10.1017/S0140525X00023992.

Buss, David M., Martie G. Haselton, Todd K. Shackelford, April L. Bleske, and Jerome C. Wakefield. "Adaptations, Exaptations, and Spandrels." *American Psychologist* 53, no. 5 (1998): 533–48. https://doi.org/10.1037/0003-066X.53.5.533.

Bütikofer, Aline, David N. Figlio, Krzysztof Karbownik, Christopher W. Kuzawa, and Kjell G. Salvanes. "Evidence That Prenatal Testosterone Transfer from Male Twins Reduces the Fertility and Socioeconomic Success of Their Female Co-Twins." *Proceedings of the National Academy of Sciences* 116, no. 14 (2019): 6749–53. https://doi.org/10.1073/pnas.1812786116.

Capel, Blanche. "Vertebrate Sex Determination: Evolutionary Plasticity of a Fundamental Switch." *Nature Reviews Genetics* 18, no. 11 (2017): 675–89. https://doi.org/10.1038/nrg.2017.60.

Caro, Shana M., Stuart A. West, and Ashleigh S. Griffin. "Sibling Conflict and Dishonest Signaling in Birds." *Proceedings of the National Academy of Sciences* 113, no. 48 (November 2016): 13803–8. https://doi.org/10.1073/pnas.1606378113.

Carter, Anthony M. "Evolution of Placental Function in Mammals: The

Molecular Basis of Gas and Nutrient Transfer, Hormone Secretion, and Immune Responses." *Physiological Reviews* 92, no. 4 (October 2012): 1543–76. https://doi.org/10.1152/physrev.00040.2011.

Carter, C. Sue, and Allison M. Perkeybile. "The Monogamy Paradox: What Do Love and Sex Have to Do With It?" *Frontiers in Ecology and Evolution* 6 (November 2018). https://www.frontiersin.org /article/10.3389/fevo.2018.00202.

Caspari, Rachel, and Sang-Hee Lee. "Older Age Becomes Common Late in Human Evolution." *Proceedings of the National Academy of Sciences* 101, no. 30 (July 2004): 10895–900. https://doi.org/10.1073 /pnas.0402857101.

Castelli, Erick C., Mateus V. de Castro, Michel S. Naslavsky, Marilia O. Scliar, Nayane S. B. Silva, Heloisa S. Andrade, Andreia S. Souza, et al. "MHC Variants Associated With Symptomatic Versus Asymptomatic SARS-CoV-2 Infection in Highly Exposed Individuals." *Frontiers in Immunology* 12 (September 2021): 742881. https://doi.org/10.3389 /fimmu.2021.742881.

Castro, Andrea, Rachel Marty Pyke, Xinlian Zhang, Wesley Kurt Thompson, Chi-Ping Day, Ludmil B. Alexandrov, Maurizio Zanetti, and Hannah Carter. "Strength of Immune Selection in Tumors Varies with Sex and Age." *Nature Communications* 11, no. 1 (2020): 4128. https://doi.org/10.1038/s41467-020-17981-0.

Cattanach, Bruce M., Colin V. Beechey, and Josephine Peters. "Interactions Between Imprinting Effects in the Mouse." *Genetics* 168, no. 1 (September 2004): 397–413. https://doi.org/10.1534/genetics.104.030064.

Cepelewicz, Jordana. "Why Nature Prefers Couples, Even for Yeast." *Quanta Magazine*, July 17, 2018. https://www.quantamagazine.org /why-nature-prefers-couples-even-for-yeast-20180717/.

Chavan, Arun Rajendra, Bhart-Anjan S. Bhullar, and Günter P. Wagner. "What Was the Ancestral Function of Decidual Stromal Cells? A Model for the Evolution of Eutherian Pregnancy." *Placenta* 40 (April 2016): 40–51. https://doi.org/10.1016/j.placenta.2016.02.012.

Chevalier-Skolnikoff, Suzanne. "Male-Female, Female-Female, and Male-Male Sexual Behavior in the Stumptail Monkey, with Special Attention to the Female Orgasm." *Archives of Sexual Behavior* 3, no. 2 (March 1974): 95–116. https://doi.org/10.1007/BF01540994.

Chuong, Edward B., Nels C. Elde, and Cédric Feschotte. "Regulatory Activities of Transposable Elements: From Conflicts to Benefits." *Nature Reviews Genetics* 18, no. 2 (2017): 71–86. https://doi.org /10.1038/nrg.2016.139.

Cloutier, Christina T., James E. Coxworth, and Kristen Hawkes. "Age-Related Decline in Ovarian Follicle Stocks Differ between Chimpanzees (*Pan troglodytes*) and Humans." *AGE* 37, no. 1 (2015): 10. https://doi.org/10.1007/s11357-015-9746-4.

Clutton-Brock, Tim, and Katherine McAuliffe. "Female Mate Choice in Mammals." *Quarterly Review of Biology* 84, no. 1 (March 2009): 3–27. https://doi.org/10.1086/596461.

Cohen, Jon. *Coming to Term: Uncovering the Truth about Miscarriage.* New York: Houghton Mifflin, 2005. https://books.google.com /books?id=Ck4YIelfRUUC.

Cohen-Bendahan, Celina C. C., Jan K. Buitelaar, Stephanie H. M. van Goozen, Jacob F. Orlebeke, and Peggy T. Cohen-Kettenis. "Is There an Effect of Prenatal Testosterone on Aggression and Other Behavioral Traits? A Study Comparing Same-Sex and Opposite-Sex Twin Girls." *Hormones and Behavior* 47, no. 2 (February 2005): 230–37. https://doi.org/10.1016/j.yhbeh.2004.10.006.

Cole, Gemma L., and John A. Endler. "Change in Male Coloration Associated with Artificial Selection on Foraging Colour Preference." *Journal of Evolutionary Biology* 31, no. 8 (August 2018): 1227–38. https://doi.org/10.1111/jeb.13300.

Colegrave, Nick. "Sex Releases the Speed Limit on Evolution." *Nature* 420, no. 6916 (2002): 664–66. https://doi.org/10.1038/nature01191.

Coleman, Seth W., Gail L. Patricelli, and Gerald Borgia. "Variable Female Preferences Drive Complex Male Displays." *Nature* 428, no. 6984 (2004): 742–45. https://doi.org/10.1038/nature02419.

Cooke, S. J., R. S. McKinley, and D. P. Philipp. "Physical Activity and Behavior of a Centrarchid Fish, *Micropterus salmoides* (Lacépède), during Spawning." *Ecology of Freshwater Fish* 10, no. 4 (December 2001): 227–37. https://doi.org/10.1034/j.1600-0633.2001.100405.x.

Cornell Lab. "House Finch Identification." All About Birds (website). Accessed July 18, 2022. https://www.allaboutbirds.org/guide/House_Finch/id.

Crespi, Bernard. "Evolutionary Medical Insights into the SARS-CoV-2

Pandemic." *Evolution, Medicine, and Public Health* 2020, no. 1 (2020): 314–22. https://doi.org/10.1093/emph/eoaa036.

———. "Oxytocin, Testosterone, and Human Social Cognition: Oxytocin and Social Behavior." *Biological Reviews* 91, no. 2 (May 2016): 390–408. https://doi.org/10.1111/brv.12175.

Critchley, Hilary O. D., Elnur Babayev, Serdar E. Bulun, Sandy Clark, Iolanda Garcia-Grau, Peter K. Gregersen, Aoife Kilcoyne, et al. "Menstruation: Science and Society." *American Journal of Obstetrics and Gynecology* 223, no. 5 (November 2020): 624–64. https://doi.org/10.1016/j.ajog.2020.06.004.

Croft, Darren P., Rufus A. Johnstone, Samuel Ellis, Stuart Nattrass, Daniel W. Franks, Lauren J. N. Brent, Sonia Mazzi, Kenneth C. Balcomb, John K. B. Ford, and Michael A. Cant. "Reproductive Conflict and the Evolution of Menopause in Killer Whales." *Current Biology* 27, no. 2 (January 2017): 298–304. https://doi.org/10.1016/j.cub.2016.12.015.

Czech, Daniel P., Joohyung Lee, Helena Sim, Clare L. Parish, Eric Vilain, and Vincent R. Harley. "The Human Testis—Determining Factor SRY Localizes in Midbrain Dopamine Neurons and Regulates Multiple Components of Catecholamine Synthesis and Metabolism." *Journal of Neurochemistry* 122, no. 2 (July 2012): 260–71. https://doi.org/10.1111/j.1471-4159.2012.07782.x.

Dale, James, Robert Montgomerie, Denise Michaud, and Peter Boag. "Frequency and Timing of Extrapair Fertilisation in the Polyandrous Red Phalarope (*Phalaropus fulicarius*)." *Behavioral Ecology and Sociobiology* 46, no. 1 (June 1999): 50–56. https://doi.org/10.1007/s002650050591.

Daly, Steven E. J., and Peter E. Hartmann. "Infant Demand and Milk Supply. Part 2: The Short-Term Control of Milk Synthesis in Lactating Women." *Journal of Human Lactation* 11, no. 1 (March 1995): 27–37. https://doi.org/10.1177/089033449501100120.

Daniloski, Zharko, Tristan X. Jordan, Hans-Hermann Wessels, Daisy A. Hoagland, Silva Kasela, Mateusz Legut, Silas Maniatis, et al. "Identification of Required Host Factors for SARS-CoV-2 Infection in Human Cells." *Cell* 184, no. 1 (January 2021): 92–105.e16. https://doi.org/10.1016/j.cell.2020.10.030.

Darwin, Charles. Darwin Correspondence Project. Cambridge University. https://www.darwinproject.ac.uk/.

———. *The Descent of Man, and Selection in Relation to Sex*. London: John Murray, 1871.

———. *On the Origin of Species by Means of Natural Selection, or Preservation of Favoured Races in the Struggle for Life*. London: John Murray, 1859.

Dávila, S. G., J. L. Campo, M. G. Gil, C. Castaño, and J. Santiago-Moreno. "Effect of the Presence of Hens on Roosters Sperm Variables." *Poultry Science* 94, no. 7 (July 2015): 1645–49. https://doi.org/10.3382/ps/pev125.

Dawkins, Richard. *The Selfish Gene*. New York: Oxford University Press, 1976.

Day, Corinne S., and Bennett G. Galef. "Pup Cannibalism: One Aspect of Maternal Behavior in Golden Hamsters." *Journal of Comparative and Physiological Psychology* 91, no. 5 (1977): 1179–89. https://doi.org/10.1037/h0077386.

Day, Felix R., Katherine S. Ruth, Deborah J. Thompson, Kathryn L. Lunetta, Natalia Pervjakova, Daniel I. Chasman, Lisette Stolk, et al. "Large-Scale Genomic Analyses Link Reproductive Aging to Hypothalamic Signaling, Breast Cancer Susceptibility and BRCA1-Mediated DNA Repair." *Nature Genetics* 47, no. 11 (September 2015): 1294–1303. https://doi.org/10.1038/ng.3412.

DeAngelis, Ross S., and Hans A. Hofmann. "Neural and Molecular Mechanisms Underlying Female Mate Choice Decisions in Vertebrates." *Journal of Experimental Biology* 223, no. 17 (September 2020): jeb207324. https://doi.org/10.1242/jeb.207324.

Dean-Jones, Lesley. "Menstrual Bleeding According to the Hippocratics and Aristotle." *Transactions of the American Philological Association* 119 (1989): 177–92. https://doi.org/10.2307/284268.

de Bruyn, P. J. N., C. A. Tosh, M. N. Bester, E. Z. Cameron, T. McIntyre, and I. S. Wilkinson. "Sex at Sea: Alternative Mating System in an Extremely Polygynous Mammal." *Animal Behaviour* 82, no. 3 (September 2011): 445–51. https://doi.org/10.1016/j.anbehav.2011.06.006.

Del Valle, Diane Marie, Seunghee Kim-Schulze, Hsin-Hui Huang, Noam D. Beckmann, Sharon Nirenberg, Bo Wang, Yonit Lavin, et al. "An

Inflammatory Cytokine Signature Predicts COVID-19 Severity and Survival." *Nature Medicine* 26, no. 10 (2020): 1636–43. https://doi.org/10.1038/s41591-020-1051-9.

Dennerstein, Lorraine, Gordon Gotts, James B. Brown, Carol A. Morse, Tim M. M. Farley, and Alain Pinol. "The Relationship between the Menstrual Cycle and Female Sexual Interest in Women with PMS Complaints and Volunteers." *Psychoneuroendocrinology* 19, no. 3 (1994): 293–304. https://doi.org/10.1016/0306-4530(94)90067-1.

Derntl, Birgit, Veronika Schopf, Kathrin Kollndorfer, and Rupert Lanzenberger. "Menstrual Cycle Phase and Duration of Oral Contraception Intake Affect Olfactory Perception." *Chemical Senses* 38, no. 1 (January 2013): 67–75. https://doi.org/10.1093/chemse/bjs084.

Diamond, Jared M. *The World until Yesterday: What Can We Learn from Traditional Societies?* New York: Viking, 2012.

Dinsdale, Natalie, Pablo Nepomnaschy, and Bernard Crespi. "The Evolutionary Biology of Endometriosis." *Evolution, Medicine, and Public Health* 9, no. 1 (2021): 174–91. https://doi.org/10.1093/emph/eoab008.

Doty, Richard L., and E. Leslie Cameron. "Sex Differences and Reproductive Hormone Influences on Human Odor Perception." *Physiology & Behavior* 97, no. 2 (May 2009): 213–28. https://doi.org/10.1016/j.physbeh.2009.02.032.

Doucet, Stéphanie M., and Robert Montgomerie. "Multiple Sexual Ornaments in Satin Bowerbirds: Ultraviolet Plumage and Bowers Signal Different Aspects of Male Quality." *Behavioral Ecology* 14, no. 4 (July 2003): 503–9. https://doi.org/10.1093/beheco/arg035.

D'Souza, Alaric W., and Günter P. Wagner. "Malignant Cancer and Invasive Placentation: A Case for Positive Pleiotropy between Endometrial and Malignancy Phenotypes." *Evolution, Medicine, and Public Health* 2014, no. 1 (2014): 136–45. https://doi.org/10.1093/emph/eou022.

Earp, Sarah E., and Donna L. Maney. "Birdsong: Is It Music to Their Ears?" *Frontiers in Evolutionary Neuroscience* 4 (November 2012):14. https://doi.org/10.3389/fnevo.2012.00014.

Easton, Judith A., Jaime C. Confer, Cari D. Goetz, and David M. Buss. "Reproduction Expediting: Sexual Motivations, Fantasies, and the Ticking Biological Clock." *Personality and Individual Differences*

49, no. 5 (October 2010): 516–20. https://doi.org/10.1016/j.paid
.2010.05.018.

Eastwick, Paul W., and Eli J. Finkel. "Sex Differences in Mate Preferences
Revisited: Do People Know What They Initially Desire in a Romantic
Partner?" *Journal of Personality and Social Psychology* 94, no. 2 (2008):
245–64. https://doi.org/10.1037/0022-3514.94.2.245.

Eastwick, Paul W., and Lucy L. Hunt. "Relational Mate Value: Consensus
and Uniqueness in Romantic Evaluations." *Journal of Personality and
Social Psychology* 106, no. 5 (2014): 728–51. https://doi.org/10.1037
/a0035884.

Elliott, John J., and Robert S. Arbib Jr. "Origin and Status of the House
Finch in the Eastern United States." *The Auk* 70, no. 1 (January 1953):
31–37. https://doi.org/10.2307/4081056.

Ellis, Samuel, Daniel W. Franks, Stuart Nattrass, Michael A. Cant,
Destiny L. Bradley, Deborah Giles, Kenneth C. Balcomb, and Darren
P. Croft. "Postreproductive Lifespans Are Rare in Mammals."
Ecology and Evolution 8, no. 5 (March 2018): 2482–94. https://
doi.org/10.1002/ece3.3856.

Ellison, Peter T., and Mary Ann Ottinger. "A Comparative Perspective
on Reproductive Aging, Reproductive Cessation, Post-
Reproductive Life, and Social Behavior." In *Sociality, Hierarchy,
Health: Comparative Biodemography: A Collection of Papers*, edited
by M. Weinstein and M. A. Lane. Washington: National Academies
Press, 2014.

Emera, Deena, Claudio Casola, Vincent J. Lynch, Derek E. Wildman,
Dalen Agnew, and Günter P. Wagner. "Convergent Evolution of
Endometrial Prolactin Expression in Primates, Mice, and Elephants
Through the Independent Recruitment of Transposable Elements."
Molecular Biology and Evolution 29, no. 1 (January 2012): 239–47.
https://doi.org/10.1093/molbev/msr189.

Emera, Deena, Roberto Romero, and Günter P. Wagner. "The Evolution
of Menstruation: A New Model for Genetic Assimilation: Explaining
Molecular Origins of Maternal Responses to Fetal Invasiveness."
BioEssays 34, no. 1 (January 2012): 26–35. https://doi.org/10.1002
/bies.201100099.

Emera, Deena, and Günter P. Wagner. "Transposable Element

Recruitments in the Mammalian Placenta: Impacts and Mechanisms." *Briefings in Functional Genomics* 11, no. 4 (July 2012): 267–76. https://doi.org/10.1093/bfgp/els013.

Emlen, Stephen T., and Lewis W. Oring. "Ecology, Sexual Selection, and the Evolution of Mating Systems." *Science* 197, no. 4300 (July 1977): 215–23. https://doi.org/10.1126/science.327754.

Emmett, Pauline M., and Imogen S. Rogers. "Properties of Human Milk and Their Relationship with Maternal Nutrition." *Early Human Development* 49, supplement (October 1997): S7–28. https://doi.org /10.1016/S0378-3782(97)00051-0.

Enard, David, and Dmitri A. Petrov. "Evidence That RNA Viruses Drove Adaptive Introgression between Neanderthals and Modern Humans." *Cell* 175, no. 2 (October 2018): 360–371.e13. https://doi.org /10.1016/j.cell.2018.08.034.

Eriksson, Nicholas, Geoffrey M. Benton, Chuong B. Do, Amy K. Kiefer, Joanna L. Mountain, David A. Hinds, Uta Francke, and Joyce Y. Tung. "Genetic Variants Associated with Breast Size Also Influence Breast Cancer Risk." *BMC Medical Genetics* 13, no. 1 (2012): 53. https:// doi.org/10.1186/1471-2350-13-53.

Exton, Michael S., Anne Bindert, Tillmann H. C. Krüger, Friedmann Scheller, Uwe Hartmann, and Manfred Schedlowski. "Cardiovascular and Endocrine Alterations After Masturbation-Induced Orgasm in Women." *Psychosomatic Medicine* 61, no. 3 (May/June 1999): 280–89. https://doi.org/10.1097/00006842-199905000-00005.

Exton, Michael S., Tillmann H. C. Krüger, Markus Koch, Erika Paulson, Wolfram Knapp, Uwe Hartmann, and Manfred Schedlowski. "Coitus-Induced Orgasm Stimulates Prolactin Secretion in Healthy Subjects." *Psychoneuroendocrinology* 26, no. 3 (April 2001): 287–94. https://doi.org/10.1016/S0306-4530(00)00053-6.

Ferris, Patrick, Bradley J. S. C. Olson, Peter L. De Hoff, Stephen Douglass, David Casero, Simon Prochnik, Sa Geng, et al. "Evolution of an Expanded Sex-Determining Locus in *Volvox*." *Science* 328, no. 5976 (April 2010): 351–54. https://doi.org/10.1126/science.1186222.

Feschotte, Cédric. "Evolutionary History and Impact of Human Transposons." *eLS* (2008). https://doi.org/10.1002/9780470015902 .a0020996.

———. "Transposable Elements and the Evolution of Regulatory Networks." *Nature Reviews Genetics* 9, no. 5 (May 2008): 397–405. https://doi.org/10.1038/nrg2337.

Finch, Caleb E., and Donna J. Holmes. "Ovarian Aging in Developmental and Evolutionary Contexts." *Annals of the New York Academy of Sciences* 1204, no. 1 (August 2010): 82–94. https://doi.org/10.1111/j.1749-6632.2010.05610.x.

Finn, C. A. "Menstruation: A Nonadaptive Consequence of Uterine Evolution." *Quarterly Review of Biology* 73, no. 2 (June 1998): 163–73. https://doi.org/10.1086/420183.

Fisher, Helen E, Arthur Aron, and Lucy L Brown. "Romantic Love: A Mammalian Brain System for Mate Choice." *Philosophical Transactions of the Royal Society B: Biological Sciences* 361, no. 1476 (December 2006): 2173–86. https://doi.org/10.1098/rstb.2006.1938.

Fisher, R. A. "The Evolution of Sexual Preference." *Eugenics Review* 7, no. 3 (1915): 184–92.

Foldes, Pierre, and Odile Buisson. "The Clitoral Complex: A Dynamic Sonographic Study." *Journal of Sexual Medicine* 6, no. 5 (May 2009): 1223–31. https://doi.org/10.1111/j.1743-6109.2009.01231.x.

Fox, C. A., H. S. Wolff, and J. A. Baker. "Measurement of Intra-Vaginal and Intra-Uterine Pressures During Human Coitus by Radio-Telemetry." *Reproduction* 22, no. 2 (July 1970): 243–51. https://doi.org/10.1530/jrf.0.0220243.

Frazer, Jennifer. "Yeast: Making Food Great for 5,000 Years. But What Exactly Is It?" *Scientific American*, September 6, 2013. https://blogs.scientificamerican.com/artful-amoeba/yeast-making-food-great-for-5000-years-but-what-exactly-is-it/.

Freese, Elisabeth Bautz, Martha I. Chu, and Ernst Freese. "Initiation of Yeast Sporulation by Partial Carbon, Nitrogen, or Phosphate Deprivation." *Journal of Bacteriology* 149, no. 3 (March 1982): 840–51. https://doi.org/10.1128/jb.149.3.840-851.1982.

Freyer, C., U. Zeller, and M. B. Renfree. "Placental Function in Two Distantly Related Marsupials." *Placenta* 28, no. 2–3 (February-March 2007): 249–57. https://doi.org/10.1016/j.placenta.2006.03.007.

Gao, Wei, Yan-Bo Sun, Wei-Wei Zhou, Zi-Jun Xiong, Luonan Chen, Hong Li, Ting-Ting Fu, et al. "Genomic and Transcriptomic Investigations of the Evolutionary Transition from Oviparity to Viviparity." *Proceedings of the National Academy of Sciences* 116, no. 9 (February 2019): 3646–55. https://doi.org/10.1073pnas.1816086116.

Garver-Apgar, Christine E., Melissa A. Eaton, Joshua M. Tybur, and Melissa Emery Thompson. "Evidence of Intralocus Sexual Conflict: Physically and Hormonally Masculine Individuals Have More Attractive Brothers Relative to Sisters." *Evolution and Human Behavior* 32, no. 6 (November 2011): 423–32. https://doi.org/10.1016/j.evolhumbehav.2011.03.005.

Geng, Sa, Peter De Hoff, and James G. Umen. "Evolution of Sexes from an Ancestral Mating-Type Specification Pathway." *PLoS Biology* 12, no. 7 (2014): e1001904. https://doi.org/10.1371/journal.pbio.1001904.

Gilbert, Scott F., and Michael J. F. Barresi. *Developmental Biology*. 11th ed. Sunderland, MA: Sinauer Associates, 2016.

Global Health 5050. "The Sex, Gender and COVID-19 Project." Accessed July 18, 2022. https://globalhealth5050.org/the-sex-gender-and-covid-19-project/.

Goldfoot, D. A., H. Westerborg-van Loon, W. Groeneveld, and A. K. Slob. "Behavioral and Physiological Evidence of Sexual Climax in the Female Stump-Tailed Macaque (*Macaca arctoides*)." *Science* 208, no. 4451 (June 1980): 1477–79. https://doi.org/10.1126/science.7384791.

Goodship, Nicola M, and Katherine L Buchanan. "Nestling Testosterone Is Associated with Begging Behaviour and Fledging Success in the Pied Flycatcher, *Ficedula hypoleuca*." *Proceedings of the Royal Society B: Biological Sciences* 273, no. 1582 (January 2006): 71–76. https://doi.org/10.1098/rspb.2005.3289.

Gorrell, Jamieson C., Andrew G. McAdam, David W. Coltman, Murray M. Humphries, and Stan Boutin. "Adopting Kin Enhances Inclusive Fitness in Asocial Red Squirrels." *Nature Communications* 1, no. 1 (June 2010): 22. https://doi.org/10.1038/ncomms1022.

Gosden, R. G., and Evelyn Telfer. "Numbers of Follicles and Oocytes in Mammalian Ovaries and Their Allometric Relationships." *Journal of Zoology* 211, no. 1 (January 1987): 169–75. https://doi.org/10.1111/j.1469-7998.1987.tb07460.x.

Gould, Stephen Jay. "Exaptation: A Crucial Tool for an Evolutionary Psychology." *Journal of Social Issues* 47, no. 3 (Fall 1991): 43–65. https://doi.org/10.1111/j.1540-4560.1991.tb01822.x.

———. "Freudian Slip." *Natural History* 96, no. 2 (1987): 14–21.

Gould, Stephen Jay, and R. C. Lewontin. "The Spandrels of San Marco and the Panglossian Paradigm: A Critique of the Adaptationist Programme." *Proceedings of the Royal Society B: Biological Sciences* 205, no. 1161 (September 1979): 581–98. https://doi.org/10.1098/rspb.1979.0086.

Grant, Peter R., and B. Rosemary Grant. "Unpredictable Evolution in a 30-Year Study of Darwin's Finches." *Science* 296, no. 5568 (April 2002): 707–11. https://doi.org/10.1126/science.1070315.

Graves, Jennifer A. Marshall. "Sex Chromosome Specialization and Degeneration in Mammals." *Cell* 124, no. 5 (March 2006): 901–14. https://doi.org/10.1016/j.cell.2006.02.024.

Griffith, Oliver W., Arun R. Chavan, Stella Protopapas, Jamie Maziarz, Roberto Romero, and Günter P. Wagner. "Embryo Implantation Evolved from an Ancestral Inflammatory Attachment Reaction." *Proceedings of the National Academy of Sciences* 114, no. 32 (July 2017): e6566–75. https://doi.org/10.1073/pnas.1701129114.

Grillot, Rachel L., Zachary L. Simmons, Aaron W. Lukaszewski, and James R. Roney. "Hormonal and Morphological Predictors of Women's Body Attractiveness." *Evolution and Human Behavior* 35, no. 3 (May 2014): 176–83. https://doi.org/10.1016/j.evolhumbehav.2014.01.001.

Grob, B., L. A. Knapp, R. D. Martin, and G. Anzenberger. "The Major Histocompatibility Complex and Mate Choice: Inbreeding Avoidance and Selection of Good Genes." *Experimental and Clinical Immunogenetics* 15, no. 3 (November 1998): 119–29. https://doi.org/10.1159/000019063.

Groothuis, Ton G. G., Wendt Müller, Nikolaus von Engelhardt, Claudio Carere, and Corine Eising. "Maternal Hormones as a Tool to Adjust Offspring Phenotype in Avian Species." *Neuroscience & Biobehavioral Reviews* 29, no. 2 (April 2005): 329–52. https://doi.org/10.1016/j.neubiorev.2004.12.002.

Guernsey, Michael W., Edward B. Chuong, Guillaume Cornelis, Marilyn B. Renfree, and Julie C. Baker. "Molecular Conservation of Marsupial and Eutherian Placentation and Lactation." *eLife* 6 (September 2017): e27450. https://doi.org/10.7554/eLife.27450.

Gunter, Jen. *The Menopause Manifesto: Own Your Health with Facts and Feminism.* Toronto: Random House, 2021.

Gurven, Michael, and Hillard Kaplan. "Longevity Among Hunter-Gatherers: A Cross-Cultural Examination." *Population and Development Review* 33, no. 2 (June 2007): 321–65. https://doi.org/10.1111/j.1728-4457.2007.00171.x.

Haig, David. "Coadaptation and Conflict, Misconception and Muddle, in the Evolution of Genomic Imprinting." *Heredity* 113, no. 2 (2014): 96–103. https://doi.org/10.1038/hdy.2013.97.

———. "Genetic Conflicts in Human Pregnancy." *Quarterly Review of Biology* 68, no. 4 (December 1993): 495–532. https://doi.org/10.1086/418300.

———. "Placental Growth Hormone-Related Proteins and Prolactin-Related Proteins." *Placenta* 29, supplement (March 2008): 36–41. https://doi.org/10.1016/j.placenta.2007.09.010.

———. "Transfers and Transitions: Parent-Offspring Conflict, Genomic Imprinting, and the Evolution of Human Life History." *Proceedings of the National Academy of Sciences* 107, supplement 1 (January 2010): 1731–35. https://doi.org/10.1073/pnas.0904111106.

Haldane, J. B. S. "A Mathematical Theory of Natural and Artificial Selection. Part I." *Transactions of the Cambridge Philosophical Society* 23 (1924): 19–41.

Haldane, J. B. S., and Edmund Brisco Ford. "The Theory of Selection for Melanism in Lepidoptera." *Proceedings of the Royal Society B: Biological Sciences* 145, no. 920 (July 1956): 303–6. https://doi.org/10.1098/rspb.1956.0038.

Hamilton, W. D. "The Genetical Evolution of Social Behaviour. II." *Journal of Theoretical Biology* 7, no. 1 (July 1964): 17–52. https://doi.org/10.1016/0022-5193(64)90039-6.

———. "The Moulding of Senescence by Natural Selection." *Journal of Theoretical Biology* 12, no. 1 (September 1966): 12–45. https://doi.org/10.1016/0022-5193(66)90184-6.

Hammer, M. L. A., and R. A. Foley. "Longevity and Life History in Hominid Evolution." *Human Evolution* 11, no. 1 (January 1996): 61–66. https://doi.org/10.1007/BF02456989.

Hao, Sha, Junli Zhao, Jianjun Zhou, Shuli Zhao, Yali Hu, and Yayi Hou.

"Modulation of 17-Estradiol on the Number and Cytotoxicity of NK Cells in Vivo Related to MCM and Activating Receptors." *International Immunopharmacology* 7, no. 13 (December 2007): 1765–75. https://doi.org/10.1016/j.intimp.2007.09.017.

Harlow, Harry F., M. K. Harlow, and S. J. Suomi. "From Thought to Therapy: Lessons from a Primate Laboratory." *American Scientist* 59, no. 5 (September-October 1971): 538–49.

Harlow, Harry F., and Robert R. Zimmermann. "Affectional Response in the Infant Monkey: Orphaned Baby Monkeys Develop a Strong and Persistent Attachment to Inanimate Surrogate Mothers." *Science* 130, no. 3373 (August 1959): 421–32. https://doi.org/10.1126/science.130.3373.421.

Harris, J. Robin. "Placental Endogenous Retrovirus (ERV): Structural, Functional, and Evolutionary Significance." *BioEssays* 20, no. 4 (April 1998): 307–16. https://doi.org/10.1002/(SICI)1521-1878(199804)20:4<307::AID-BIES7>3.0.CO;2-M.

Haselton, Martie G., Mina Mortezaie, Elizabeth G. Pillsworth, April Bleske-Rechek, and David A. Frederick. "Ovulatory Shifts in Human Female Ornamentation: Near Ovulation, Women Dress to Impress." *Hormones and Behavior* 51, no. 1 (January 2007): 40–45. https://doi.org/10.1016/j.yhbeh.2006.07.007.

Hawkes, K., J. F. O'Connell, N. G. B. Jones, H. Alvarez, and E. L. Charnov. "Grandmothering, Menopause, and the Evolution of Human Life Histories." *Proceedings of the National Academy of Sciences* 95, no. 3 (February 1998): 1336–39. https://doi.org/10.1073/pnas.95.3.1336.

Healthily. "How many eggs does a woman have?" December 12, 2021. https://www.livehealthily.com/self-care/women-eggs.

Hill, Geoffrey E. "Geographic Variation in Male Ornamentation and Female Mate Preference in the House Finch: A Comparative Test of Models of Sexual Selection." *Behavioral Ecology* 5, no. 1 (Spring 1994): 64–73. https://doi.org/10.1093/beheco/5.1.64.

———. "Male Mate Choice and the Evolution of Female Plumage Coloration in the House Finch." *Evolution* 47, no. 5 (October 1993): 1515–25. https://doi.org/10.1111/j.1558-5646.1993.tb02172.x.

Hill, Geoffrey E., and Kevin McGraw. "Mate Attentiveness, Seasonal Timing of Breeding and Long-Term Pair Bonding in the House

Finch (*Carpodacus mexicanus*)." *Behaviour* 141, no. 1 (January 2004): 1–13. https://doi.org/10.1163/156853904772746565.

Hillis, David. "To Egg or Not to Egg: That Is the Evolutionary Question." *Biodiversity Blog*. March 6, 2019. https://biodiversity.utexas.edu /news/entry/to-egg-or-not-to-egg.

Hinde, Katherine. "Richer Milk for Sons but More Milk for Daughters: Sex-Biased Investment during Lactation Varies with Maternal Life History in Rhesus Macaques." *American Journal of Human Biology* 21, no. 4 (July/August 2009): 512–19. https://doi.org/10.1002/ajhb.20917.

Hoag, N., J. R. Keast, and H. E. O'Connell. "The 'G-Spot' Is Not a Structure Evident on Macroscopic Anatomic Dissection of the Vaginal Wall." *Journal of Sexual Medicine* 14, no. 2 (February 2017): e32. https://doi.org /10.1016/j.jsxm.2016.12.079.

Hollis, David M., Marvy V. Price, Richard W. Hill, David W. Hall, Marta J. Laskowski. Principles of Life. 3rd ed. New York: W. H. Freeman and Company, 2019.

Hrdy, Sarah Blaffer. *Mother Nature: Maternal Instincts and How They Shape the Human Species*. New York: Ballantine Books, 2000.

———. "Of Marmosets, Men, and the Transformative Power of Babies." In *Costly and Cute: How Helpless Newborns Made Us Human*, edited by Wenda R. Trevathan and Karen R. Rosenberg, 177–204. Albuquerque: University of New Mexico Press, 2016.

———. "Variable Postpartum Responsiveness among Humans and Other Primates with 'Cooperative Breeding': A Comparative and Evolutionary Perspective." *Hormones and Behavior* 77 (January 2016): 272–83. https://doi.org/10.1016/j.yhbeh.2015.10.016.

Hrdy, Sarah Blaffer, and Judith M. Burkart. "The Emergence of Emotionally Modern Humans: Implications for Language and Learning." *Philosophical Transactions of the Royal Society B: Biological Sciences* 375, no. 1803 (July 2020): 20190499. https://doi.org/10.1098 /rstb.2019.0499.

Huber, Susanne, and Martin Fieder. "Evidence for a Maximum 'Shelf-Life' of Oocytes in Mammals Suggests That Human Menopause May Be an Implication of Meiotic Arrest." *Scientific Reports* 8, no. 1 (September 2018): 14099. https://doi.org/10.1038/s41598-018-32502-2.

interACT. "Media Guide: Covering the Intersex Community." interACT:

Advocates for Intersex Youth. Uploaded January 2017. https://interactadvocates.org/wp-content/uploads/2017/01/INTERSEX-MEDIAGUIDE-interACT.pdf.

Irving, Aaron T., Matae Ahn, Geraldine Goh, Danielle E. Anderson, and Lin-Fa Wang. "Lessons from the Host Defences of Bats, a Unique Viral Reservoir." *Nature* 589, no. 7842 (January 2021): 363–70. https://doi.org/10.1038/s41586-020-03128-0.

Jarvis, Gavin E. "Early Embryo Mortality in Natural Human Reproduction: What the Data Say." *F1000Research* 5 (November 2016): 2765. https://doi.org/10.12688/f1000research.8937.1.

Jasieńska, Grazyna, Anna Ziomkiewicz, Peter T. Ellison, Susan F. Lipson, and Inger Thune. "Large Breasts and Narrow Waists Indicate High Reproductive Potential in Women." *Proceedings of the Royal Society B: Biological Sciences* 271, no. 1545 (June 2004): 1213–17. https://doi.org/10.1098/rspb.2004.2712.

Javed, Asma, and Aida Lteif. "Development of the Human Breast." *Seminars in Plastic Surgery* 27, no. 01 (2013): 005–012. https://doi.org/10.1055/s-0033-1343989.

Johnco, Carly, Ladd Wheeler, and Alan Taylor. "They Do Get Prettier at Closing Time: A Repeated Measures Study of the Closing-Time Effect and Alcohol." *Social Influence* 5, no. 4 (2010): 261–71. https://doi.org/10.1080/15534510.2010.487650.

Kaneko-Ishino, Tomoko, and Fumitoshi Ishino. "Retrotransposon Silencing by DNA Methylation Contributed to the Evolution of Placentation and Genomic Imprinting in Mammals: Evolution of Placenta and Imprinting." *Development, Growth & Differentiation* 52, no. 6 (August 2010): 533–43. https://doi.org/10.1111/j.1440-169X.2010.01194.x.

Kaplan, Helen Singer. *Disorders of Sexual Desire and Other New Concepts and Techniques in Sex Therapy.* New York: Simon and Schuster, 1979.

Katchman, Benjamin. "Cellular Fountain of Youth." ASU - Ask A Biologist. June 7, 2012. https://askabiologist.asu.edu/plosable/cellular-fountain-youth.

Kawazu, Isao, Masakatsu Kino, Konomi Maeda, Yasuhiro Yamaguchi, and Yutaka Sawamukai. "Induction of Oviposition by the Administration of Oxytocin in Hawksbill Turtles." *Zoological*

Science 31, no. 12 (December 2014): 831–35. https://doi.org/10.2108/zs140032.

Kelsey, T. W., R. A. Anderson, P. Wright, S. M. Nelson, and W. H. B. Wallace. "Data-Driven Assessment of the Human Ovarian Reserve." *Molecular Human Reproduction* 18, no. 2 (February 2012): 79–87. https://doi.org/10.1093/molehr/gar059.

Kerr, Jeffrey B., Michelle Myers, and Richard A. Anderson. "The Dynamics of the Primordial Follicle Reserve." *Reproduction* 146, no. 6 (2013): R205–15. https://doi.org/10.1530/REP-13-0181.

Kim, Pilyoung, Lane Strathearn, and James E. Swain. "The Maternal Brain and Its Plasticity in Humans." *Hormones and Behavior* 77 (January 2016): 113–23. https://doi.org/10.1016/j.yhbeh.2015.08.001.

Kim, Young, Christopher A. Latz, Charles S. DeCarlo, Sujin Lee, C. Y. Maximilian Png, Pavel Kibrik, Eric Sung, Olamide Alabi, and Anahita Dua. "Relationship between Blood Type and Outcomes Following COVID-19 Infection." *Seminars in Vascular Surgery* 34, no. 3 (September 2021): 125–31. https://doi.org/10.1053/j.semvascsurg.2021.05.005.

Kobelt, Georg Ludwig. *Die männlichen und weiblichen Wollust-Organe des Menschen und einiger Säugetiere.* Freiburg: Adolph Emmerling. 1844.

Komisaruk, Barry R., Carlos Beyer, and Beverly Whipple. "Orgasm." *The Psychologist* 21 (February 2008): 100–103.

Konner, Melvin. *Childhood.* Boston: Little, Brown, 1991.

Kościński, Krzysztof, Rafał Makarewicz, and Zbigniew Bartoszewicz. "Stereotypical and Actual Associations of Breast Size with Mating-Relevant Traits." *Archives of Sexual Behavior* 49, no. 3 (2020): 821–36 .https://doi.org/10.1007/s10508-019-1464-z.

Kowalewski, Mariusz P., Aykut Gram, Ewa Kautz, and Felix R. Graubner. "The Dog: Nonconformist, Not Only in Maternal Recognition Signaling." In *Regulation of Implantation and Establishment of Pregnancy in Mammals*, edited by Rodney D. Geisert and Fuller W. Bazer, 216:215–37. Cham: Springer International, 2015. https://doi.org/10.1007/978-3-319-15856-3_11.

Kraemer, Sebastian. "The Fragile Male." *BMJ* 321, no. 7276 (December 2000): 1609–12. https://doi.org/10.1136/bmj.321.7276.1609.

Krakauer, Alan H. "Kin Selection and Cooperative Courtship in Wild

Turkeys." *Nature* 434, no. 7029 (March 2005): 69–72. https://doi.org/10.1038/nature03325.

Kruger, Daniel J., and Randolph M. Nesse. "An Evolutionary Life-History Framework for Understanding Sex Differences in Human Mortality Rates." *Human Nature* 17, no. 1 (March 2006): 74–97. https://doi.org/10.1007/s12110-006-1021-z.

Kshitiz, Junaid Afzal, Jamie D. Maziarz, Archer Hamidzadeh, Cong Liang, Eric M. Erkenbrack, Hong Nam Kim, et al. "Evolution of Placental Invasion and Cancer Metastasis Are Causally Linked." *Nature Ecology & Evolution* 3, no. 12 (November 2019): 1743–53. https://doi.org/10.1038/s41559-019-1046-4.

Kunz, Thomas H., and David J. Hosken. "Male Lactation: Why, Why Not and Is It Care?" *Trends in Ecology & Evolution* 24, no. 2 (February 2009): 80–85. https://doi.org/10.1016/j.tree.2008.09.009.

Kuzawa, Christopher W., Harry T. Chugani, Lawrence I. Grossman, Leonard Lipovich, Otto Muzik, Patrick R. Hof, Derek E. Wildman, Chet C. Sherwood, William R. Leonard, and Nicholas Lange. "Metabolic Costs and Evolutionary Implications of Human Brain Development." *Proceedings of the National Academy of Sciences* 111, no. 36 (August 2014): 13010–15. https://doi.org/10.1073/pnas.1323099111.

Lahdenperä, Mirkka, Duncan O. S. Gillespie, Virpi Lummaa, and Andrew F. Russell. "Severe Intergenerational Reproductive Conflict and the Evolution of Menopause." *Ecology Letters* 15, no. 11 (November 2012): 1283–90. https://doi.org/10.1111/j.1461-0248.2012.01851.x.

Laisk, Triin, Olga Tšuiko, Tatjana Jatsenko, Peeter Hõrak, Marjut Otala, Mirkka Lahdenperä, Virpi Lummaa, Timo Tuuri, Andres Salumets, and Juha S Tapanainen. "Demographic and Evolutionary Trends in Ovarian Function and Aging." *Human Reproduction Update* 25, no. 1 (January-February 2019): 34–50. https://doi.org/10.1093/humupd/dmy031.

Lamm, Melissa S., Hui Liu, Neil J. Gemmell, and John R. Godwin. "The Need for Speed: Neuroendocrine Regulation of Socially-Controlled Sex Change." *Integrative and Comparative Biology* 55, no. 2 (August 2015): 307–22. https://doi.org/10.1093/icb/icv041.

Leonard, William R. "Lifestyle, Diet, and Disease: Comparative Perspectives on the Determinants of Chronic Health Risks." In *Evolution in Health and Disease*, edited by Stephen C. Stearns and

Jacob C. Koella, 265–76. Oxford: Oxford University Press, 2007. https://doi.org/10.1093/acprof:oso/9780199207466.003.0020.

Lewis, C. S. *Letters of C. S. Lewis*. Edited by W. H. Lewis and Walter Hooper. London: Collins, 1988.

Lipson, S. F., and P. T. Ellison. "Comparison of Salivary Steroid Profiles in Naturally Occurring Conception and Non-Conception Cycles." *Human Reproduction* 11, no. 10 (October 1996): 2090–96. https://doi.org/10.1093/oxfordjournals.humrep.a019055.

Lloyd, Elisabeth A. *The Case of the Female Orgasm: Bias in the Science of Evolution*. Cambridge, MA: Harvard University Press, 2005.

Lloyd, Elisabeth A., and Stephen Jay Gould. "Exaptation Revisited: Changes Imposed by Evolutionary Psychologists and Behavioral Biologists." *Biological Theory* 12, no. 1 (January 2017): 50–65. https://doi.org/10.1007/s13752-016-0258-y.

Loehlin, John C., and Nicholas G. Martin. "Dimensions of Psychological Masculinity-Femininity in Adult Twins from Opposite-Sex and Same-Sex Pairs." *Behavior Genetics* 30, no. 1 (January 2000): 19–28. https://doi.org/10.1023/a:1002082325784.

Loffredo, Christopher, and Gerald Borgia. "Male Courtship Vocalizations as Cues for Mate Choice in the Satin Bowerbird (*Ptilonorhynchus violaceus*)." *American Naturalist* 128 (December 1986): 773–94. https://doi.org/10.1093/auk/103.1.189.

Loke, Y. W. *Life's Vital Link: The Astonishing Role of the Placenta*. Oxford: Oxford University Press, 2013.

Lynch, Kathleen S., A. Stanley Rand, Michael J. Ryan, and Walter Wilczynski. "Plasticity in Female Mate Choice Associated with Changing Reproductive States." *Animal Behaviour* 69, no. 3 (March 2005): 689–99. https://doi.org/10.1016/j.anbehav.2004.05.016.

Lynch, Vincent J., Robert D. Leclerc, Gemma May, and Günter P. Wagner. "Transposon-Mediated Rewiring of Gene Regulatory Networks Contributed to the Evolution of Pregnancy in Mammals." *Nature Genetics* 43, no. 11 (September 2011): 1154–59. https://doi.org/10.1038/ng.917.

Lynch, Vincent J., Mauris C. Nnamani, Aurélie Kapusta, Kathryn Brayer, Silvia L. Plaza, Erik C. Mazur, Deena Emera, et al. "Ancient Transposable Elements Transformed the Uterine Regulatory

Landscape and Transcriptome during the Evolution of Mammalian Pregnancy." *Cell Reports* 10, no. 4 (February 2015): 551–61. https://doi.org/10.1016/j.celrep.2014.12.052.

Lyon, Bruce E., and Daizaburo Shizuka. "Extreme Offspring Ornamentation in American Coots Is Favored by Selection within Families, Not Benefits to Conspecific Brood Parasites." *Proceedings of the National Academy of Sciences* 117, no. 4 (2020): 2056–64. https://doi.org/10.1073/pnas.1913615117.

Macy, Icie G., Helen A. Hunscher, Eva Donelson, and Betty Nims. "Human Milk Flow." *American Journal of Diseases of Children* 39, no. 6 (June 1930): 1186–1204. https://doi.org/10.1001/archpedi.1930.01930180036004.

Madlon-Kay, Diane J. "'Witch's Milk': Galactorrhea in the Newborn." *American Journal of Diseases of Children* 140, no. 3 (March 1986): 252. https://doi.org/10.1001/archpedi.1986.02140170078035.

Maestripieri, Dario, and Kelly A. Carroll. "Causes and Consequences of Infant Abuse and Neglect in Monkeys." *Aggression and Violent Behavior* 5, no. 3 (May-June 2000): 245–54. https://doi.org/10.1016/S1359-1789(98)00019-6.

Majerus, Michael E. N. "Industrial Melanism in the Peppered Moth, *Biston betularia*: An Excellent Teaching Example of Darwinian Evolution in Action." *Evolution: Education and Outreach* 2, no. 1 (2009): 63–74. https://doi.org/10.1007/s12052-008-0107-y.

Manire, Charles A., Lynne Byrd, Corie L. Therrien, and Kelly Martin. "Mating-Induced Ovulation in Loggerhead Sea Turtles, *Caretta caretta*." *Zoo Biology* 27, no. 3 (May/June 2008): 213–25. https://doi.org/10.1002/zoo.20171.

Manning, C. Jo, Edward K. Wakeland, and Wayne K. Potts. "Communal Nesting Patterns in Mice Implicate MHC Genes in Kin Recognition." *Nature* 360, no. 6404 (December 1992): 581–83. https://doi.org/10.1038/360581a0.

Martin, Robert D. "The Evolution of Human Reproduction: A Primatological Perspective." *American Journal of Physical Anthropology* 134, no. S45 (2007): 59–84. https://doi.org/10.1002/ajpa.20734.

———. "Human Reproduction: A Comparative Background for Medical Hypotheses." *Journal of Reproductive Immunology* 59, no. 2 (August 2003): 111–35. https://doi.org/10.1016/S0165-0378(03)00042-1.

Masters, William H., Virginia E. Johnson, and Reproductive Biology Research Foundation. *Human Sexual Response*. New York: Bantam Books, 1966.

Mauvais-Jarvis, Franck, Noel Bairey Merz, Peter J. Barnes, Roberta D. Brinton, Juan-Jesus Carrero, Dawn L. DeMeo, Geert J. De Vries, et al. "Sex and Gender: Modifiers of Health, Disease, and Medicine." *The Lancet* 396, no. 10250 (August 2020): 565–82. https://doi.org /10.1016/S0140-6736(20)31561-0.

May, Robin C. "Gender, Immunity and the Regulation of Longevity." *BioEssays* 29, no. 8 (August 2007): 795–802. https://doi.org/10.1002 /bies.20614.

McCarthy, Margaret. "Sex Differences in the Developing Brain as a Source of Inherent Risk." *Dialogues in Clinical Neuroscience* 18, no. 4 (2016): 361–72. https://doi.org/10.31887/DCNS.2016.18.4 /mmccarthy.

McClellan, Holly L., Susan J. Miller, and Peter E. Hartmann. "Evolution of Lactation: Nutrition *v.* Protection with Special Reference to Five Mammalian Species." *Nutrition Research Reviews* 21, no. 2 (2008): 97–116. https://doi.org/10.1017/S0954422408100749.

McGovern, Patrick E. *Uncorking the Past: The Quest for Wine, Beer, and Other Alcoholic Beverages*. Berkeley: University of California Press, 2009.

McLaughlin, Jessica E. "Menstrual Cycle." Merck Manual. Merck & Co, Inc. Last modified September 2022. https://www.merckmanuals.com/home /women-s-health-issues/biology-of-the-female-reproductive-system /menstrual-cycle.

Medawar, Peter Brian. *An Unsolved Problem of Biology*. London: H. K. Lewis, 1952.

Mennill, Daniel J., Alexander V. Badyaev, Leslie M. Jonart, and Geoffrey E. Hill. "Male House Finches with Elaborate Songs Have Higher Reproductive Performance." *Ethology* 112, no. 2 (February 2006): 174–80. https://doi.org/10.1111/j.1439-0310.2006.01145.x.

Moffett, Ashley, and Charlie Loke. "Immunology of Placentation in Eutherian Mammals." *Nature Reviews Immunology* 6, no. 8 (August 2006): 584–94. https://doi.org/10.1038/nri1897.

Morandini, V., and M. Ferrer. "Sibling Aggression and Brood Reduction:

A Review." *Ethology Ecology & Evolution* 27, no. 1 (2015): 2–16. https://doi.org/10.1080/03949370.2014.880161.

Morgan, Elaine. *The Aquatic Ape Hypothesis.* London: Souvenir, 2017.

Morris, Desmond. *The Naked Ape.* New York: Random House, 1999.

Mosconi, Lisa. *The XX Brain: The Groundbreaking Science Empowering Women to Maximize Cognitive Health and Prevent Alzheimer's Disease.* New York: Avery, 2020.

Mosconi, Lisa, Valentina Berti, Crystal Quinn, Pauline McHugh, Gabriella Petrongolo, Ricardo S. Osorio, Christopher Connaughty, et al. "Perimenopause and Emergence of an Alzheimer's Bioenergetic Phenotype in Brain and Periphery." *PLOS ONE* 13, no. 2 (February 2018): e0193314. https://doi.org/10.1371/journal.pone.0193314.

Muehlenbein, Michael P., ed. *Human Evolutionary Biology.* Cambridge UK: Cambridge University Press, 2010.

Murray, Carson M., Margaret A. Stanton, Elizabeth V. Lonsdorf, Emily E. Wroblewski, and Anne E. Pusey. "Chimpanzee Fathers Bias Their Behaviour towards Their Offspring." *Royal Society Open Science* 3, no. 11 (November 2016): 160441. https://doi.org/10.1098/rsos.160441.

Natterson-Horowitz, Barbara, and Kathryn Bowers. *Zoobiquity: What Animals Can Teach Us about Health and the Science of Healing.* New York: Alfred A. Knopf, 2012.

Nattrass, Stuart, Darren P. Croft, Samuel Ellis, Michael A. Cant, Michael N. Weiss, Brianna M. Wright, Eva Stredulinsky, et al. "Postreproductive Killer Whale Grandmothers Improve the Survival of Their Grandoffspring." *Proceedings of the National Academy of Sciences* 116, no. 52 (December 2019): 26669–73. https://doi.org/10.1073/pnas.1903844116.

Nielsen, Sofie Holtsmark, Lucy van Dorp, Charlotte J. Houldcroft, Anders G. Pedersen, Morten E. Allentoft, Lasse Vinner, Ashot Margaryan, et al. "31,600-Year-Old Human Virus Genomes Support a Pleistocene Origin for Common Childhood Infections." Preprint, submitted June 28, 2021. https://doi.org/10.1101/2021.06.28.450199.

Ober, Carole, Lowell R. Weitkamp, Nancy Cox, Harvey Dytch, Donna Kostyu, and Sherman Elias. "HLA and Mate Choice in Humans." *American Journal of Human Genetics* 61, no. 3 (1997): 497–504. https://doi.org/10.1086/515511.

O'Connell, Helen E., and John O. L. DeLancey. "Clitoral Anatomy in Nulliparous, Healthy, Premenopausal Volunteers Using Unenhanced Magnetic Resonance Imaging." *Journal of Urology* 173, no. 6 (June 2005): 2060–63. https://doi.org/10.1097/01.ju.0000158446.21396.c0.

O'Connell, Lauren A., and Hans A. Hofmann. "The Vertebrate Mesolimbic Reward System and Social Behavior Network: A Comparative Synthesis." *Journal of Comparative Neurology* 519, no. 18 (December 2011): 3599–639. https://doi.org/10.1002/cne.22735.

Oftedal, Olav T. "The Evolution of Lactation in Mammalian Species." In *Milk, Mucosal Immunity and the Microbiome: Impact on the Neonate*, edited by Pearay L. Ogra, W. Allan Walker, and Bo Lönnerdal, Nestlé Nutrition Institute Workshop Series, 94:1–10. Basel: Karger, 2020. https://doi.org/10.1159/000505577.

———. "The Evolution of Milk Secretion and Its Ancient Origins." *Animal* 6, no. 3 (2012): 355–68. https://doi.org/10.1017/S1751731111001935.

———. "The Mammary Gland and Its Origin During Synapsid Evolution." *Journal of Mammary Gland Biology and Neoplasia* 7, no. 3 (July 2002): 225–52. https://doi.org/10.1023/A:1022896515287.

Oh, Kevin P., and Alexander V. Badyaev. "Structure of Social Networks in a Passerine Bird: Consequences for Sexual Selection and the Evolution of Mating Strategies." *American Naturalist* 176, no. 3 (September 2010): E80–89. https://doi.org/10.1086/655216.

Olival, Kevin J., Parviez R. Hosseini, Carlos Zambrana-Torrelio, Noam Ross, Tiffany L. Bogich, and Peter Daszak. "Host and Viral Traits Predict Zoonotic Spillover from Mammals." *Nature* 546, no. 7660 (June 2017): 646–50. https://doi.org/10.1038/nature22975.

Opie, Christopher, Quentin D. Atkinson, Robin I. M. Dunbar, and Susanne Shultz. "Male Infanticide Leads to Social Monogamy in Primates." *Proceedings of the National Academy of Sciences* 110, no. 33 (July 2013): 13328–32. https://doi.org/10.1073/pnas.1307903110.

Orbach, Dara N., Christopher D. Marshall, Bernd Würsig, and Sarah L. Mesnick. "Variation in Female Reproductive Tract Morphology of the Common Bottlenose Dolphin (*T ursiops truncatus*)." *Anatomical Record* 299, no. 4 (April 2016): 520–37. https://doi.org/10.1002/ar.23318.

Page, Abigail E., and Jennifer C. French. "Reconstructing Prehistoric Demography: What Role for Extant Hunter-Gatherers?" *Evolutionary*

Anthropology: Issues, News, and Reviews 29, no. 6 (November/December 2020): 332–45. https://doi.org/10.1002/evan.21869.

Pairo-Castineira, Erola, Sara Clohisey, Lucija Klaric, Andrew D. Bretherick, Konrad Rawlik, Dorota Pasko, Susan Walker, et al. "Genetic Mechanisms of Critical Illness in COVID-19." *Nature* 591, no. 7848 (December 2021): 92–98. https://doi.org/10.1038/s41586-020-03065-y.

Papper, Zack, Natalie M. Jameson, Roberto Romero, Amy L. Weckle, Pooja Mittal, Kurt Benirschke, Joaquin Santolaya-Forgas, et al. "Ancient Origin of Placental Expression in the Growth Hormone Genes of Anthropoid Primates." *Proceedings of the National Academy of Sciences* 106, no. 40 (October 2009): 17083–88. https://doi.org/10.1073/pnas.0908377106.

Paschou, Stavroula *A.*, Panagiotis Anagnostis, Dimitra I. Pavlou, Andromachi Vryonidou, Dimitrios G. Goulis, and Irene Lambrinoudaki. "Diabetes in Menopause: Risks and Management." *Current Vascular Pharmacology* 17, no. 6 (2019): 556–63. https://doi.org/10.2174/1570161116666180625124405.

Patricelli, Gail L., Seth W. Coleman, and Gerald Borgia. "Male Satin Bowerbirds, *Ptilonorhynchus violaceus*, Adjust Their Display Intensity in Response to Female Startling: An Experiment with Robotic Females." *Animal Behaviour* 71, no. 1 (January 2006): 49–59. https://doi.org/10.1016/j.anbehav.2005.03.029.

Pauls, Rachel N. "Anatomy of the Clitoris and the Female Sexual Response: Clitoral Anatomy and Sexual Function." *Clinical Anatomy* 28, no. 3 (April 2015): 376–84. https://doi.org/10.1002/ca.22524.

Pavlicev, Mihaela, and Errol R. Norwitz. "Human Parturition: Nothing More Than a Delayed Menstruation." *Reproductive Sciences* 25, no. 2 (February 2018): 166–73. https://doi.org/10.1177/1933719117725830.

Pavlicev, Mihaela, and Günter Wagner. "The Evolutionary Origin of Female Orgasm: Evolution of Female Orgasm." *Journal of Experimental Zoology Part B: Molecular and Developmental Evolution* 326, no. 6 (September 2016): 326–37. https://doi.org/10.1002/jez.b.22690.

Pavlicev, Mihaela, Andreja Moset Zupan, Amanda Barry, Savannah Walters, Kristin M. Milano, Harvey J. Kliman, and Günter P. Wagner. "An Experimental Test of the Ovulatory Homolog

Model of Female Orgasm." *Proceedings of the National Academy of Sciences* 116, no. 41 (September 2019): 20267–73. https://doi .org/10.1073/pnas.1910295116.

Penn, Dustin J., and Wayne Potts. "How Do Major Histocompatibility Complex Genes Influence Odor and Mating Preferences." *Advances in Immunology* 69 (1998): 411–36. https://doi.org/10.1016 /s0065-2776(08)60612-4.

Penn, Dustin J., and Ken R. Smith. "Differential Fitness Costs of Reproduction between the Sexes." *Proceedings of the National Academy of Sciences* 104, no. 2 (January 2007): 553–58. https://doi.org /10.1073/pnas.0609301103.

Pennebaker, James W., Mary Anne Dyer, R. Scott Caulkins, Debra Lynn Litowitz, Phillip L. Ackerman, Douglas B. Anderson, and Kevin M. McGraw. "Don't the Girls' Get Prettier at Closing Time: A Country and Western Application to Psychology." *Personality and Social Psychology Bulletin* 5, no. 1 (January 1979): 122–25. https://doi.org /10.1177/014616727900500127.

Peper, Jiska S., and Ronald E. Dahl. "The Teenage Brain: Surging Hormones—Brain-Behavior Interactions During Puberty." *Current Directions in Psychological Science* 22, no. 2 (April 2013): 134–39. https://doi.org/10.1177/0963721412473755.

Perls, Thomas T., and Ruth C. Fretts. "The Evolution of Menopause and Human Life Span." *Annals of Human Biology* 28, no. 3 (2001): 237–45. https://doi.org/10.1080/030144601300119052.

Peters, Jo. "The Role of Genomic Imprinting in Biology and Disease: An Expanding View." *Nature Reviews Genetics* 15, no. 8 (June 2014): 517–30. https://doi.org/10.1038/nrg3766.

Pfaus, James G., Tina Scardochio, Mayte Parada, Christine Gerson, Gonzalo R. Quintana, and Genaro A. Coria-Avila. "Do Rats Have Orgasms?" *Socioaffective Neuroscience & Psychology* 6, no. 1 (2016): 31883. https://doi.org/10.3402/snp.v6.31883.

Phoenix, Charles H., Robert W. Goy, Arnold A. Gerall, and William C. Young. "Organizing Action of Prenatally Administered Testosterone Propionate on the Tissues Mediating Mating Behavior in the Female Guinea Pig." *Endocrinology* 65, no. 3 (September 1959): 369–82. https://doi.org/10.1210/endo-65-3-369.

Place, Ned J., Alexandra M. Prado, Mariela Faykoo-Martinez, Miguel Angel Brieño-Enriquez, David F. Albertini, and Melissa M. Holmes. "Germ Cell Nests in Adult Ovaries and an Unusually Large Ovarian Reserve in the Naked Mole-Rat." *Reproduction* 161, no. 1 (January 2021): 89–98. https://doi.org/10.1530/REP-20-0304.

Potts, Malcolm, and R. V. Short. *Ever since Adam and Eve: The Evolution of Human Sexuality.* Cambridge UK: Cambridge University Press, 1999.

Potts, Wayne K. "Wisdom through Immunogenetics." *Nature Genetics* 30, no. 2 (February 2002): 130–31. https://doi.org/10.1038/ng0202-130.

Prentice, Andrew, Alison Paul, Ann Prentice, Alison Black, Tim Cole, and Roger Whitehead. "Cross-Cultural Differences in Lactational Performance." In *Human Lactation 2*, edited by Margrit Hamosh and Armond S. Goldman, 13–44. Boston: Springer, 1986.

Profet, Margie. "Menstruation as a Defense Against Pathogens Transported by Sperm." *Quarterly Review of Biology* 68, no. 3 (September 1993): 335–86. https://doi.org/10.1086/418170.

Qiao, Huanyu, H. B. D. Prasada Rao, Yan Yun, Sumit Sandhu, Jared H. Fong, Manali Sapre, Michael Nguyen, et al. "Impeding DNA Break Repair Enables Oocyte Quality Control." *Molecular Cell* 72, no. 2 (October 2018): 211–221.e3. https://doi.org/10.1016/j.molcel.2018.08.031.

Rao, Sheila, and Janelle S. Ayres. "Resistance and Tolerance Defenses in Cancer: Lessons from Infectious Diseases." *Seminars in Immunology* 32 (August 2017): 54–61. https://doi.org/10.1016/j.smim.2017.08.004.

Rebar, Darren, Nathan W. Bailey, Benjamin J. M. Jarrett, and Rebecca M. Kilner. "An Evolutionary Switch from Sibling Rivalry to Sibling Cooperation, Caused by a Sustained Loss of Parental Care." *Proceedings of the National Academy of Sciences* 117, no. 5 (January 2020): 2544–50. https://doi.org/10.1073/pnas.1911677117.

Regan, Jennifer C., and Linda Partridge. "Gender and Longevity: Why Do Men Die Earlier than Women? Comparative and Experimental Evidence." *Best Practice & Research Clinical Endocrinology & Metabolism* 27, no. 4 (August 2013): 467–79. https://doi.org/10.1016/j.beem.2013.05.016.

Renfree, Marilyn B., Shunsuke Suzuki, and Tomoko Kaneko-Ishino. 2013. "The Origin and Evolution of Genomic Imprinting and Viviparity in Mammals." *Philosophical Transactions of the Royal Society B: Biological*

Sciences 368, no. 1609 (January 2013): 20120151. https://doi.org /10.1098/rstb.2012.0151.

Rettew, Jennifer A., Yvette M. Huet-Hudson, and Ian Marriott. "Testosterone Reduces Macrophage Expression in the Mouse of Toll-Like Receptor 4, a Trigger for Inflammation and Innate Immunity." *Biology of Reproduction* 78, no. 3 (March 2008): 432–37. https://doi.org/10.1095/biolreprod.107.063545.

Reusch, Thorsten B. H., Michael A. Häberli, Peter B. Aeschlimann, and Manfred Milinski. "Female Sticklebacks Count Alleles in a Strategy of Sexual Selection Explaining MHC Polymorphism." *Nature* 414, no. 6861 (November 2001): 300–302. https://doi.org/10.1038/35104547.

Rilling, James K. "The Neural and Hormonal Bases of Human Parental Care." *Neuropsychologia* 51, no. 4 (March 2013): 731–47. https://doi.org /10.1016/j.neuropsychologia.2012.12.017.

Robinson, Dionne P., Maria E. Lorenzo, William Jian, and Sabra L. Klein. "Elevated 17-Estradiol Protects Females from Influenza A Virus Pathogenesis by Suppressing Inflammatory Responses." *PLoS Pathogens* 7, no. 7 (July 2011): e1002149. https://doi.org/10.1371 /journal.ppat.1002149.

Robinson, Richard. "For Mammals, Loss of Yolk and Gain of Milk Went Hand in Hand." *PLoS Biology* 6, no. 3 (March 2008): e77. https://doi.org /10.1371/journal.pbio.0060077.

Rodd, F. Helen, Kimberly A. Hughes, Gregory F. Grether, and Colette T. Baril. "A Possible Non-Sexual Origin of Mate Preference: Are Male Guppies Mimicking Fruit?" *Proceedings of the Royal Society B: Biological Sciences* 269, no. 1490 (March 2002): 475–81. https:// doi.org/10.1098/rspb.2001.1891.

Ronald, Kelly L., Esteban Fernández-Juricic, and Jeffrey R. Lucas. "Mate Choice in the Eye and Ear of the Beholder? Female Multimodal Sensory Configuration Influences Her Preferences." *Proceedings of the Royal Society B: Biological Sciences* 285, no. 1878 (May 2018): 20180713. https://doi.org/10.1098/rspb.2018.0713.

Roney, James R., and Zachary L. Simmons. "Hormonal Predictors of Sexual Motivation in Natural Menstrual Cycles." *Hormones and Behavior* 63, no. 4 (April 2013): 636–45. https://doi.org/10.1016 /j.yhbeh.2013.02.013.

Rudolph, Marion, Wolf-Dietrich Döcke, Andrea Müller, Astrid Menning, Lars Röse, Thomas Matthias Zollner, and Isabella Gashaw. "Induction of Overt Menstruation in Intact Mice." *PLoS ONE* 7, no. 3 (March 2012): e32922. https://doi.org/10.1371/journal .pone.0032922.

Ruth, K.S., Day, F.R., Hussain, J. et al. "Genetic insights into biological mechanisms governing human ovarian ageing." *Nature* 596, 393–397 (2021). https://doi.org/10.1038/s41586-021-03779-7.

Ryan, Michael J. "Darwin, Sexual Selection, and the Brain." *Proceedings of the National Academy of Sciences* 118, no. 8 (February 2021): e2008194118. https://doi.org/10.1073/pnas.2008194118.

——. "Fickle Females?" *Nature* 428, no. 6984 (April 2004): 708–9. https://doi.org/10.1038/428708a.

——. *A Taste for the Beautiful: The Evolution of Attraction.* Princeton, NJ: Princeton University Press, 2018.

Saini, Angela. *Inferior: How Science Got Women Wrong and the New Research That's Rewriting the Story.* Boston: Beacon Press, 2017.

Salker, Madhuri, Gijs Teklenburg, Mariam Molokhia, Stuart Lavery, Geoffrey Trew, Tepchongchit Aojanepong, Helen J. Mardon, et al. "Natural Selection of Human Embryos: Impaired Decidualization of Endometrium Disables Embryo-Maternal Interactions and Causes Recurrent Pregnancy Loss." *PLoS ONE* 5, no. 4 (April 2010): e10287. https://doi.org/10.1371/journal.pone.0010287.

Samson, Michel, Frédérick Libert, Benjamin J. Doranz, Joseph Rucker, Corinne Liesnard, Claire-Michèle Farber, Sentob Saragosti, et al. "Resistance to HIV-1 Infection in Caucasian Individuals Bearing Mutant Alleles of the CCR-5 Chemokine Receptor Gene." *Nature* 382, no. 6593 (August 1996): 722–25. https://doi.org /10.1038/382722a0.

Sanger, Thomas J., Marissa L. Gredler, and Martin J. Cohn. "Resurrecting Embryos of the Tuatara, *Sphenodon punctatus*, to Resolve Vertebrate Phallus Evolution." *Biology Letters* 11, no. 10 (October 2015): 20150694. https://doi.org/10.1098/rsbl.2015.0694.

Schiebinger, Londa. "Why Mammals Are Called Mammals: Gender

Politics in Eighteenth-Century Natural History." *American Historical Review* 98, no. 2 (April 1993): 382. https://doi.org/10.2307/2166840.

Scully, Eileen P., Jenna Haverfield, Rebecca L. Ursin, Cara Tannenbaum, and Sabra L. Klein. "Considering How Biological Sex Impacts Immune Responses and COVID-19 Outcomes." *Nature Reviews Immunology* 20, no. 7 (June 2020): 442–47. https://doi.org/10.1038/s41577-020-0348-8.

Sear, Rebecca, and Ruth Mace. "Who Keeps Children Alive? A Review of the Effects of Kin on Child Survival." *Evolution and Human Behavior* 29, no. 1 (January 2008): 1–18. https://doi.org/10.1016/j.evolhumbehav.2007.10.001.

Setchell, Joanna M. "Do Female Mandrills Prefer Brightly Colored Males?" *International Journal of Primatology* 26, no. 4 (August 2005): 715–35. https://doi.org/10.1007/s10764-005-5305-7.

———. "Sexual Selection and the Differences between the Sexes in Mandrills (*M andrillus sphinx*)." *American Journal of Physical Anthropology* 159, no. S61 (January 2016): 105–29. https://doi.org/10.1002/ajpa.22904.

Shubin, Neil. *Your Inner Fish: A Journey into the 3.5-Billion-Year History of the Human Body*. New York: Vintage Books, 2009.

Singh, Prim B., Richard E. Brown, and Bruce Roser. "MHC Antigens in Urine as Olfactory Recognition Cues." *Nature* 327, no. 6118 (May 1987): 161–64. https://doi.org/10.1038/327161a0.

Slepicka, Priscila Ferreira, Amritha Varshini Hanasoge Somasundara, and Camila O. dos Santos. "The Molecular Basis of Mammary Gland Development and Epithelial Differentiation." *Seminars in Cell & Developmental Biology* 114 (June 2021): 93–112. https://doi.org/10.1016/j.semcdb.2020.09.014.

Small, Meredith F. *Our Babies, Ourselves: How Biology and Culture Shape the Way We Parent*. New York: Anchor Books, 1999.

Smith, John Maynard. "Survival by Suicide." *New Scientist* 67 (August 1975): 496–97.

Smith, Ken R., Heidi A. Hanson, Geraldine P. Mineau, and Saundra S. Buys. "Effects of *BRCA1* and *BRCA2* Mutations on Female Fertility."

Proceedings of the Royal Society B: Biological Sciences 279, no. 1732 (October 2012): 1389–95. https://doi.org/10.1098/rspb.2011.1697.

Speroff, Leon, and Marc A. Fritz. *Clinical Gynecologic Endocrinology and Infertility*. 7th ed. Philadelphia: Lippincott Williams & Wilkins, 2005.

Spigler, R. B., K. S. Lewers, D. S. Main, and T.-L. Ashman. "Genetic Mapping of Sex Determination in a Wild Strawberry, *Fragaria virginiana*, Reveals Earliest Form of Sex Chromosome." *Heredity* 101, no. 6 (September 2008): 507–17. https://doi.org/10.1038/hdy.2008.100.

Stansfield, F. J., J. O. Nöthling, and W. R. Allen. "The Progression of Small-Follicle Reserves in the Ovaries of Wild African Elephants (*Loxodonta africana*) from Puberty to Reproductive Senescence." *Reproduction, Fertility and Development* 25, no. 8 (2013): 1165. https://doi.org/10.1071/RD12296.

Stearns, Stephen C. "Evolutionary Medicine: Its Scope, Interest and Potential." *Proceedings of the Royal Society B: Biological Sciences* 279, no. 1746 (November 2012): 4305–21. https://doi.org/10.1098/rspb.2012.1326.

———. "Frontiers in Molecular Evolutionary Medicine." *Journal of Molecular Evolution* 88, no. 1 (2020): 3–11. https://doi.org/10.1007/s00239-019-09893-5.

Stearns, Stephen C., and Dieter Ebert. "Evolution in Health and Disease: Work in Progress." *Quarterly Review of Biology* 76, no. 4 (December 2001): 417–32. https://doi.org/10.1086/420539.

Storey, Anne E., and Toni E. Ziegler. "Primate Paternal Care: Interactions between Biology and Social Experience." *Hormones and Behavior* 77 (January 2016): 260–71. https://doi.org/10.1016/j.yhbeh.2015.07.024.

Strassmann, Beverly I. "The Evolution of Endometrial Cycles and Menstruation." *Quarterly Review of Biology* 71, no. 2 (June 1996): 181–220. https://doi.org/10.1086/419369.

———. "Menstrual Cycling and Breast Cancer: An Evolutionary Perspective." *Journal of Women's Health* 8, no. 2 (March 1999): 193–202. https://doi.org/10.1089/jwh.1999.8.193.

Symons, Donald. *The Evolution of Human Sexuality*. Oxford: Oxford University Press, 1979.

Tecle, Eillen, Hector Sequoyah Reynoso, Ruixuan Wang, and Pascal Gagneux. "The Female Reproductive Tract Contains Multiple Innate Sialic Acid-Binding Immunoglobulin-like Lectins (Siglecs) That Facilitate Sperm Survival." *Journal of Biological Chemistry* 294, no. 31 (August 2019): 11910–19. https://doi.org/10.1074/jbc.RA119.008729.

Teklenburg, Gijs, Madhuri Salker, Mariam Molokhia, Stuart Lavery, Geoffrey Trew, Tepchongchit Aojanepong, Helen J. Mardon, et al. "Natural Selection of Human Embryos: Decidualizing Endometrial Stromal Cells Serve as Sensors of Embryo Quality upon Implantation." *PLoS ONE* 5, no. 4 (April 2010): e10258. https://doi.org/10.1371/journal.pone.0010258.

Temmerman, M., H. Verstraelen, G. Martens, and A. Bekaert. "Delayed Childbearing and Maternal Mortality." *European Journal of Obstetrics & Gynecology and Reproductive Biology* 114, no. 1 (May 2004): 19–22. https://doi.org/10.1016/j.ejogrb.2003.09.019.

Thomas, David L., Chloe L. Thio, Maureen P. Martin, Ying Qi, Dongliang Ge, Colm O'hUigin, Judith Kidd, et al. "Genetic Variation in *IL28B* and Spontaneous Clearance of Hepatitis C Virus." *Nature* 461, no. 7265 (October 2009): 798–801. https://doi.org/10.1038/nature08463.

Thornhill, Randy. "Alternative Female Choice Tactics in the Scorpionfly *Hylobittacus apicalis* (Mecoptera) and Their Implications." *American Zoologist* 24, no. 2 (May 1984): 367–83. https://doi.org/10.1093/icb/24.2.367.

Torgerson, David J., Ruth E. Thomas, and David M. Reid. "Mothers and Daughters Menopausal Ages: Is There a Link?" *European Journal of Obstetrics & Gynecology and Reproductive Biology* 74, no. 1 (July 1997): 63–66. https://doi.org/10.1016/S0301-2115(97)00085-7.

Trevathan, Wenda. *Ancient Bodies, Modern Lives: How Evolution Has Shaped Women's Health.* New York: Oxford University Press, 2010.

Trivers, Robert L. "Parent-Offspring Conflict." *American Zoologist* 14, no. 1 (February 1974): 249–64. https://doi.org/10.1093/icb/14.1.249.

Troisi, Alfonso, and Monica Carosi. "Female Orgasm Rate Increases with Male Dominance in Japanese Macaques." *Animal Behaviour* 56, no. 5 (November 1998): 1261–66. https://doi.org/10.10006/anbe.1998.0898.

Tuppi, Marcel, Sebastian Kehrloesser, Daniel W. Coutandin, Valerio Rossi, Laura M. Luh, Alexander Strubel, Katharina Hötte, et al. "Oocyte DNA Damage Quality Control Requires Consecutive Interplay of CHK2 and CK1 to Activate P63." *Nature Structural & Molecular Biology* 25, no. 3 (February 2018): 261–69. https://doi.org/10.1038/s41594-018-0035-7.

Uy, J. Albert C., Gail L. Patricelli, and Gerald Borgia. "Complex Mate Searching in the Satin Bowerbird *Ptilonorhynchus violaceus*." *American Naturalist* 158, no. 5 (November 2001): 530–42. https://doi.org/10.1086/323118.

Varyani, Fumi, John O. Fleming, and Rick M. Maizels. "Helminths in the Gastrointestinal Tract as Modulators of Immunity and Pathology." *American Journal of Physiology-Gastrointestinal and Liver Physiology* 312, no. 6 (June 2017): G537–49. https://doi.org/10.1152/ajpgi.00024.2017.

Vasey, Paul L., and Hester Jiskoot. "The Biogeography and Evolution of Female Homosexual Behavior in Japanese Macaques." *Archives of Sexual Behavior* 39, no. 6 (2010): 1439–41. https://doi.org/10.1007/s10508-009-9518-2.

Vitzthum, Virginia J. "Evolutionary Models of Women's Reproductive Functioning." *Annual Review of Anthropology* 37, no. 1 (2008): 53–73. https://doi.org/10.1146/annurev.anthro.37.081407.085112.

Viveiros, Anissa, Jaslyn Rasmuson, Jennie Vu, Sharon L. Mulvagh, Cindy Y. Y. Yip, Colleen M. Norris, and Gavin Y. Oudit. "Sex Differences in COVID-19: Candidate Pathways, Genetics of ACE2, and Sex Hormones." *American Journal of Physiology-Heart and Circulatory Physiology* 320, no. 1 (January 2021): H296–304. https://doi.org/10.1152/ajpheart.00755.2020.

Wagner, Günter P., Kshitiz, Anasuya Dighe, and Andre Levchenko. "The Coevolution of Placentation and Cancer." *Annual Review of Animal Biosciences* 10, no. 1 (February 2022): 259–79. https://doi.org/10.1146/annurev-animal-020420-031544.

Wagner, Günter P., and Mihaela Pavlicev. "Origin, Function, and Effects of Female Orgasm: All Three Are Different." *Journal of Experimental*

Zoology Part B: Molecular and Developmental Evolution 328, no. 4 (June 2017): 299–303. https://doi.org/10.1002/jez.b.22737.

Wagner, Günter P., Yingchun Tong, Deena Emera, and Roberto Romero. "An Evolutionary Test of the Isoform Switching Hypothesis of Functional Progesterone Withdrawal for Parturition: Humans Have a Weaker Repressive Effect of PR-A than Mice." *Journal of Perinatal Medicine* 40, no. 4 (2012): 345–51. https://doi.org/10.1515/jpm-2011-0256.

Watson, Christine J., and Walid T. Khaled. "Mammary Development in the Embryo and Adult: A Journey of Morphogenesis and Commitment." *Development* 135, no. 6 (March 2008): 995–1003. https://doi.org/10.1242/dev.005439.

Wedekind, C., T. Seebeck, F. Bettens, and A. J. Paepke. "MHC-Dependent Mate Preferences in Humans." *Proceedings of the Royal Society B: Biological Sciences* 260, no. 1359 (June 1995): 245–49. https://doi.org/10.1098/rspb.1995.0087.

Wilkinson, Philip, and Steve Noon. *A Child through Time: The Book of Children's History*. New York: DK Publishing, 2017.

Williams, George C. "Pleiotropy, Natural Selection, and the Evolution of Senescence." *Evolution* 11, no. 4 (December 1957): 398–411. https://doi.org/10.2307/2406060.

Winterbottom, M., T. Burke, and T. R. Birkhead. "A Stimulatory Phalloid Organ in a Weaver Bird." *Nature* 399, no. 6731 (May 1999): 28. https://doi.org/10.1038/19884.

Wolf, Alli. *Glitoris*. 2017. Sculpture. Sydney, Australia.

Wooding, Peter, and Graham Burton. "Fish, Amphibian, Bird and Reptile Placentation." Chap. 3 in *Comparative Placentation*. Berlin: Springer Berlin Heidelberg, 2008. https://doi.org/10.1007/978-3-540-78797-6_3.

World Health Organization Task Force on Adolescent Reproductive Health. "World Health Organization Multicenter Study on Menstrual and Ovulatory Patterns in Adolescent Girls. II. Longitudinal Study of Menstrual Patterns in the Early Postmenarcheal Period, Duration of Bleeding Episodes and Menstrual Cycles." *Journal of Adolescent Health Care* 7, no. 4 (July 1986): 236–44. https://doi.org/10.1016/S0197-0070(86)80015-8.

Worst-Online-Dater. "Tinder Experiments II: Guys, Unless You Are Really Hot You Are Probably Better off Not Wasting Your Time on Tinder—a Quantitative Socio-Economic Study." Medium, March 24, 2015. https://medium.com/@worstonlinedater/tinder-experiments-ii-guys-unless-you-are-really-hot-you-are-probably-better-off-not-wasting-your-2ddf370a6e9a.

Xiong, Jing, Seong Su Kang, Zhihao Wang, Xia Liu, Tan-Chun Kuo, Funda Korkmaz, Ashley Padilla, et al. "FSH Blockade Improves Cognition in Mice with Alzheimer's Disease." Nature 603, no. 7901 (March 2022): 470–476. https://doi.org/10.1038/s41586-022-04463-0.

Yamazaki, K., E. A. Boyse, V. Mike, H. T. Thaler, B. J. Mathieson, J. Abbott, J. Boyse, Z. A. Zayas, and L. Thomas. "Control of Mating Preferences in Mice by Genes in the Major Histocompatibility Complex." Journal of Experimental Medicine 144, no. 5 (November 1976): 1324–35. https://doi.org/10.1084%2Fjem.144.5.1324.

Zhang, Qian, Paul Bastard, Zhiyong Liu, Jérémie Le Pen, Marcela Moncada-Velez, Jie Chen, Masato Ogishi, et al. "Inborn Errors of Type I IFN Immunity in Patients with Life-Threatening COVID-19." Science 370, no. 6515 (September 2020): eabd4570. https://doi.org/10.1126/science.abd4570.

Zietsch, Brendan P., and Pekka Santtila. "No Direct Relationship between Human Female Orgasm Rate and Number of Offspring." Animal Behaviour 86, no. 2 (August 2013): 253–55. https://doi.org/10.1016/j.anbehav.2013.05.011.

Index

About the Author

DEENA EMERA, PhD, is an evolutionary geneticist, writer, and educator. She earned a bachelor's degree from UC Berkeley, a master's degree from NYU, and a PhD from Yale. She currently serves as a scientist and writer-in-residence at the world-renowned Buck Institute's Center for Reproductive Long -evity and Equality. She lives in the San Francisco Bay Area with her family.

© Lisa Keating